A QUESTION OF CHEMISTRY

A QUESTION
OF CHEMISTRY

CREATIVE PROBLEMS FOR
CRITICAL THINKERS

John Garratt, Tina Overton & Terry Threlfall

Pearson Education Limited

Edinburgh Gate, Harlow

Essex CM20 2JE, England

and Associated Companies throughout the world

© Pearson Education Limited 1999

The right of John Garratt, Tina Overton and Terry Threlfall to be identified as authors of this work has been asserted by them in accordance with the Copyright, Designs and Patents Act 1988

First published 1999

British Library Cataloguing in Publication Data

A catalogue entry for this title is available from the British Library

ISBN 0-582-29838-5

Transferred to digital print on demand, 2005

Set by 35 in Monotype Garamond 11/14 and Univers
Produced by Longman Singapore (Pte)
Printed & bound by Antony Rowe Ltd, Eastbourne

CONTENTS

INTRODUCTION

The idea behind this book is simple. We wanted our students to learn *how* to think and not just *what* to think; we wanted them to recognise that learning to be a chemist involves more than learning chemical facts and that chemistry is not (as we have heard it described) 'a purely factual subject'; we wanted them to learn to question, to think critically and creatively, and to make judgements. We realised that most of the problems available require students to find the best (standard) method by which to arrive at the (already known) correct answer. This is the kind of problem which reinforces the idea that learning chemistry is a matter of learning facts and following known procedures. Bodner's idea of problem solving is quite different and reflects more closely our understanding of what chemists need to be able to do:

Problem solving is what you do when you don't know what to do; otherwise it's not a problem.[1]

We realised that, in order to give our students the opportunity to learn *how* to think, we need other kinds of problems or exercises. We started to devise some for ourselves and this book is the result.

The notion that education involves learning how to think is as old as the notion of education itself: we claim no originality in wanting to put the idea into practice. Haldane put it like this:

It is the sole purpose of the university teacher to induce people to think.[2]

Haldane himself, being a quintessentially creative thinker, had his own ideas of what it means to think like a scientist. A description which appeals to us is Feynman's

Science is a way to teach . . . how to handle doubt and uncertainty, what the rules of evidence are, how to think about things so that judgements can be made . . .[3]

This was a comment made around 1950 in irritation at students who 'disappointed him by meekly refusing to ask questions'. We do not know whether Feynman's teaching methods overcame the problem, but we do know that he was not alone in his concern about students not thinking for themselves. Nearly 20 years later, de Bono expressed his concern that education (and not just science education) was much more concerned with memorising fact than with learning to think.

. . . the long years of education are mostly concerned with knowledge. Fact is piled upon fact and little if anzy time is spent thinking. . . . On the whole it must be more important to be skilled in thinking than to be stuffed with facts.[4]

We fear that little has changed since then. But we were encouraged that an important part of de Bono's philosophy is that it is possible to *learn* how to think effectively.

However, de Bono's publications did not provide us with the materials we wanted. We wanted our students to use their *chemistry* to learn to think. We see this as having three main advantages: it is the best way of learning to think like a professional chemist, it reinforces and puts into perspective their chemical knowledge, and it provides a motivation for them to enjoy their chemistry.

A turning point in our own thinking was our discovery of the MENO Thinking Skills Service of the University of Cambridge. This organisation has defined a set of skills which underlie the ability to think. They have used this to devise a set of tests through which to assess the potential to think, in the belief that this would be a good guide to an individual's potential to develop in any academic discipline. The MENO exercises, because they are intended as assessment exercises, must be concerned with general knowledge (be context free) and must have an unambiguously correct answer.

What attracted us to them was the potential to use the same style of exercise as a learning experience for undergraduate chemists. To play this role they needed to be based in chemistry (or at least in science) and there would be a positive advantage if (at least some) exercises had no unambiguously correct answer.

The exercises in this book grew from our discussions of the MENO exercises.

Two types of exercises 'understanding an argument' and 'constructing an argument' (Sections 1 and 2) are based directly on two of the MENO exercises. The others, 'critical reading', 'making judgements', and 'reference trails' are based on our own experiences of teaching, and seem to us to be a natural extension of the MENO exercises.

The exercises in 'understanding an argument' are designed to encourage students to evaluate arguments which are presented in short passages. Because this is, for most students, an unfamiliar task, we have simplified it by offering a selection of statements from which they choose the most appropriate one. A key point is that most of the statements we offer are correct, but may not be relevant to the evaluation of the argument. This encourages students to recognise that a correct statement is not necessarily the 'right' answer. The intention is that by completing these simplified exercises, students will learn to evaluate arguments without the prompts provided here.

In 'constructing an argument' students are asked to reassemble an argument from three randomised sentences. In most of these examples there is more than one logical sequence of sentences. The preferred choice is often governed by the depth of understanding of the chemistry in question (which is always simple) or by the approach to science. In many of these exercises it is especially rewarding to try to analyse the chain of reasoning which has led to a particular choice; it can reveal interesting differences between those who think of chemistry as a process and those who regard it as knowledge to be remembered.

'Critical reading' is a natural extension of 'understanding an argument'. The main difference is that the passages on which these exercises are based are all taken from a published work (textbook, journal or magazine). The objective is not to criticise the passage, but to recognise any assumptions which the writer is making about the context or about the reader's knowledge. Some of these exercises involve the interpretation of data. Others extend into exercises concerned with writing, a skill which is increasingly important for chemists. They need to be able to communicate the nature of their subject to others who are not chemists, or not specialists in a particular area of chemistry. The failure to do this effectively in the past is a major reason why the public image of chemistry is so poor. (We have not included more extensive exercises in writing because plenty exist already. Examples can be found in *Communicating Chemistry* by P D Bailey available from the Royal Society of Chemistry and in *53 Interesting Communication Exercises for Science Students* by S Habeshaw and D Steeds published by Technical and Education Services Ltd.)

In the section 'making judgements' we exploit more completely the principle of using questions to which there are several appropriate answers. These range from problems arising from the ambiguity in the meaning of familiar words, to the need to make decisions about experimental procedures.

The section on 'reference trails' is the only one for which all exercises have unambiguously correct answers. These exercises have a place here because students need to develop the skill and judgement to assimilate the key information in a scientific paper, and use this important resource as a source of specific information. We have restricted these reference trails to printed papers, and not extended them to searching the Internet. The Internet is a good source of material for discussion and for critical evaluation (since there is no quality control on items put there). However, a key feature of the web is that what is available changes all the time, so that tasks we might suggest now may be out of date before this book is published. This is not true of published papers where the skill of finding what one wants does not depend on whether the papers concerned were published yesterday or ten years ago. In that sense our reference trails will not go out of date.

In spite of the care we have taken, we would be surprised if everyone is satisfied with all of our exercises and commentaries. However it may still be possible to use defective exercises to encourage thought and discussion, which is the main purpose of the book.

THE STRUCTURE OF THE BOOK

Each of the first five sections is devoted to one of the five types of exercise. Each section starts with a short explanation of how to tackle them. This is followed by a varying number of exercises. We have made some attempt to put what we judge to be the easiest ones early in the chapter.

We have not divided the materials according to sub-disciplines of organic, inorganic, physical, analytical and biological chemistry. A powerful reason for not doing so is our belief that the division of science into disciplines and sub-disciplines is essentially arbitrary and artificial; this is reflected in the difficulty of unequivocal assignment of several of our exercises to a particular sub-discipline. Another reason is that the chemical knowledge required is very simple – much of it is covered in most pre-university courses and all of it should be within the grasp of any academic teaching at first- or second-year university level. However, the index is arranged to help you to find exercises which fall within particular topics.

In the commentaries we provide our preferred solutions, usually with comments, on exercises with an odd number. This seemed to us the best compromise between two extremes. We preferred not to provide a full set of commentaries which might encourage readers to avoid the thinking stage by going straight to our solution. On the other hand, our experience is that most users welcome some guidance.

HOW TO USE THIS BOOK

This book asks students to think like scientists, to recognise that the most interesting questions can be approached in different ways and have no single correct answer, and to see their tutors as facilitators of learning rather than as providers of information. It is a book which engages students in active learning and encourages them to talk to each other about chemistry. This is what our friends and colleagues who have tried our exercises for themselves tell us happens.

When we started to use our materials with students (and with willing academic colleagues) we were surprised and delighted to observe that they managed to achieve much more than our primary objective of using chemistry to introduce thinking skills. These are activities that require and encourage active participation and engagement with the material and this seems to generate enthusiasm. Sometimes students (and even tutors) initially feel uncomfortable with this kind of

exercise; in a teaching situation they are not used to facing problems where there is no well-practiced route to an answer. If you feel this way we urge you to persevere; in our experience students quickly recognise the need to engage actively with the material.

These exercises have been trialled by colleagues in several different institutions; their feedback has been overwhelmingly enthusiastic. They describe the novelty of hearing their students arguing about chemistry and of students continuing their deliberations outside the classroom. The exercises seem to be very effective at promoting communication and cooperative learning and most students feel able to make an individual worthwhile contribution – probably because there are often several effective ways of tackling them and more than one good answer so that students need not have the fear of being wrong.

We suggest that you start with the more straightforward ones at the beginning of each chapter. The commentary shows how we have tackled every other one. Alternatively, you can look through the exercises to find some which are close to your own particular interests. These are the ones with which you are most likely to feel comfortable, and will give you a model for tackling others.

An important aspect of solving these problems is the students' own analysis of how they approached the problem concerned and how they formulated their response. By doing this they begin to understand something of their own thinking processes and develop new strategies for solving problems. This helps them to develop those personal and professional skills now recognised as being an important part of higher education.

The exercises can be used in a variety of ways; we do not believe that there is a correct or even a best way. The book can be used by individual students interested in working alone and in developing their ideas about scientific argument and the process of science. The exercises can also be used in workshops or classes with a tutor present. Those in Sections 1, 2 and 4 are particularly suitable for students working in small groups to reach solutions which they then explain to the whole class and justify. We have also used them to provide 'buzz' sessions to break a lecture, and as one amongst a collection of more conventional tasks within tutorials. Many of the exercises in Section 3 (Critical reading) include passages which may take a few minutes to read and understand. The most effective discussions of these occur after individuals have reached their own conclusions.

The activities in Section 5 (Reference trails) provide essentially no opportunity for collaborative work, and so are best worked on alone.

We do not doubt that imaginative tutors will find many ways of using this book and the ideas behind it in their teaching.

REFERENCES

1. G Bodner, *J. Chem. Ed.*, 1986, **63**, 873
2. J B S Haldane, *Daedalus*, Kegan Paul, Trench, Trubner, 1924
3. J Gleik, *Genius: Richard Feynman and Modern Physics*, Little, Brown, 1992
4. E de Bono, *The Five Day Course in Thinking*, Pengiun, 1967

ACKNOWLEDGEMENTS

Any faults and mistakes are entirely our own responsibility. Nevertheless we are grateful to many friends and colleagues who have both provided us with suggestions for exercises (Joe McGinnis, University of Teesside; Stephen Breuer, University of Lancaster; Pat Bailey, Heriot-Watt University), and checked and criticised individual exercises for us (particularly Terry Kee, Leeds; Andrew Horn, York; Paul Walton, York; Doug Clow, York; and Stephen Breuer, Lancaster). We are grateful to Alex Johnstone, Glasgow, and Michael Byrne, Northumbria, for encouraging us to extend and develop our collection of exercises into a publishable state, and to numerous colleagues for stimulating discussions; these include three project students, Alison Loftus, Mandy Holbrook and Mark Gifford, whose project work has made a valuable contribution to our thinking.

The book is enlivened by the creative thinking of our illustrator, Doug Lawrence. The manuscript has been typed and retyped many times with astonishing speed and accuracy by Barbara Jones without whose word-processing skills we would never have completed the project.

Finally, we must thank the many students (at Hull) who have struggled with early versions and whose enthusiastic response to the problems encouraged us to develop them further.

UNDERSTANDING AN ARGUMENT

In this section an 'argument' does not refer to a 'discussion involving disagreement' but to 'a chain of reasoning'. Chemists must be able to make a critical evaluation of such chains of reasoning. The exercises in this section are designed to help you to develop this skill. Each exercise starts with a short passage which is usually an argument, but may (as in Exercise 1) be a description of a situation. Each passage is accompanied by a choice of additional statements one of which has a particular bearing on the argument. You are asked to identify the one which, in your opinion, best meets the criterion given. For example, you may be asked to identify the statement which:

* Most strengthens the argument.
* Best identifies the flaw or weakness in the argument.
* Most weakens the argument.
* Best states an underlying (unspoken) assumption in the argument.

Do pay careful attention to the criterion which you are asked to use to select one of the statements.

Critical evaluation of an argument involves being able to explain your conclusions to others. An important part of these exercises is therefore the justification of your choice of the alternatives given. This means that you must be able to explain to others not only why you regard your choice as satisfactory, but also why you prefer it to the alternatives offered.

When you have completed a number of these exercises, you should develop the confidence to evaluate an argument without the prompts which are provided by the accompanying statements we have provided. In order to get started, we have provided a commentary on the first exercise. In Section 6 you will find commentaries on all the rest of the odd-numbered exercises.

1 | An analyst is asked to determine the proportion by weight of iron in a sample of ground up powder. The proposed method involves dissolving a known quantity of the powder in concentrated HNO_3, making the solution up to a known volume, and using atomic absorption spectrometry to determine the concentration of iron in the solution. The analyst carries out the procedure as described, but

is disturbed to find that a small amount of residue will not dissolve in the nitric acid.

Which one of the following additional pieces of information does the analyst need in order to calculate the proportion of iron in the sample?

A The mass of the insoluble residue.

B That the insoluble residue does not contain any iron.

C The proportion of the total sample which is insoluble.

D The mass of material obtained after evaporating a sample of known volume of the solution to dryness.

GETTING STARTED

Our choice is B.

The amount of iron extracted from the sample can easily be calculated from the concentration in the known volume of the solution. If B is known to be true, then this amount is the total amount of iron in the original sample. Unless insoluble material forms a protective layer around the iron, it would be expected that all the iron would be dissolved by the treatment with concentrated HNO_3. Nevertheless, it would be difficult to prove that the insoluble residue contained no iron.

Neither A nor C can give useful information if B is true. However, if B is not true, the analyst has a real problem, and must know either A or C, and also the proportion by weight of iron in the residual solid. In practice, if only a tiny proportion of the original solid failed to dissolve, then the amount of iron it may contain will be negligible.

Under no circumstances can D give useful information about the residual solid, since material extracted from the original sample may be lost as vapour, and nitrate ions from the nitric acid may form part of the dried solid.

2 The pH of an aqueous solution is approximately equal to minus $\log_{10}$ of the hydrogen ion concentration. An acid such as hydrochloric acid is completely dissociated in water. So if 0.1 mol of HCl is dissolved in 1 dm³ of water the concentration of hydrogen ions is 0.1 mol dm⁻³ and the solution will have a pH of almost exactly 1. Similarly, if the solution is diluted to 1/10 of the concentration, the pH will be close to 2 and if it is diluted to 1 in 10^6 of the original solution the pH will be close to 7.

Which one of the following is the best statement of the flaw in this argument?

A HCl is not 100% dissociated in aqueous solution.

B pH should be defined in terms of the activity of the hydrogen ion.

c HCl is not the only source of hydrogen ions in the system.

D The hydrogen ion does not exist as such in aqueous solution; it is better represented as H_3O^+.

E A solution containing 10^{-7} mol dm^{-3} of Cl$^-$ is so dilute as to have an unmeasurable effect on pH.

3 Several times a year crowds gather in the cathedral at Naples to 'witness the miraculous liquefaction of the blood of St Janarius'. The crowd are shown a phial which is half full of a red substance which looks like blood. During the ceremony the phial is repeatedly turned upside down to see if it has liquefied. Liquifaction may take minutes, hours or even days and will occur at any time of year, summer or winter.

Which one of the following provides the best hypothesis to explain this phenomenon?

A This is a genuine miracle.

B The contents of the phial are photosensitive and liquefy on exposure to light.

C Liquefaction and solidification of the contents of the phial results from a periodic growth of a micro-organism.

D The substance is thixotropic, changing from solid to liquid when mechanically disturbed.

E The substance is a hygroscopic, deliquescent solid, becoming liquid when it absorbs moisture from the air.

4 The systematic name for a branched chain alkane is based on the longest chain. Side chains are given the name of the appropriate alkyl group and prefixed by a number which indicates the carbon atom in the longest chain to which the side chain is attached. As a general rule numbering starts from the end of the parent chain which will give the lowest number. The systematic name for the compound shown below is therefore 2,4-dimethyl-5-ethylhexane

$$\begin{array}{c} CH_3 \\ | \\ CH_2 \\ | \\ H_3C-CH-\underset{H_2}{C}-CH-\underset{H}{C}-CH_3 \\ \quad\ \ | \qquad\ \ | \\ \quad\ \ CH_3 \quad\ \ CH_3 \end{array}$$

Which one of the following statements best expresses the flaw in this argument?

A   There is no flaw in the argument.
B   The correct name is 2-ethyl-3,5-dimethylhexane.
C   The correct name is 2,4,5-trimethylheptane.
D   The correct name is 3,4,6-trimethylheptane.

5 | The equilibrium constant for the reaction

$$CH_3CO_2H + C_2H_5OH \rightleftharpoons CH_3CO_2C_2H_5 + H_2O$$

is close to 1. Therefore, when ethanol and ethanoic acid react together, the resulting reaction mixture always contains a mixture of reactants and products. Since the boiling points of ethanol and ethyl ethanoate are respectively 77 °C and 78.5 °C, it is difficult to separate ethanol from ethyl ethanoate by distillation. It is therefore not worth trying to synthesise the ester from the acid and the alcohol.

Which one of the following is the best statement of the flaw in the argument?

A   Modern separation techniques ought to make it possible to separate all esters from their parent alcohols.
B   Suitable catalysts favour the formation of esters.
C   Using excess acid it is possible to convert almost all the alcohol into ester, and this can easily be separated from the excess acid.
D   Esters can be readily synthesised by the reaction between alcohols and acid anhydrides, and this reaction goes virtually to completion.

6 | A molybdenum atom (group 6) has six valence electrons. Carbon monoxide may act as a two-electron donor when coordinated to a transition metal atom as a ligand. Hence the total number of valence electrons on the Mo atom in Mo(CO)$_7$ would be 6 + (7 × 2) = 20. Mo(CO)$_7$ is therefore unlikely to be stable.

Which one of the following is an underlying assumption of the argument above?

A   The compound Mo(CO)$_6$, with exactly 18 valence electrons, is very stable.
B   Organometallic compounds of transition metals with more than 18 valence electrons on the central metal are rarely stable.
C   CO ligands are π acids.

D 6-coordinate compounds are more stable than 7-coordinate compounds.

E $Mo(CO)_7$ cannot be an octahedral compound.

7 | All of the original helium and hydrogen that was present in the Earth's atmosphere when the Earth was originally formed has been lost. Nevertheless, some helium is detectable in the Earth's atmosphere. The small amount of helium that is present in today's atmosphere is therefore the product of ongoing radioactive decay.

Which one of the following statements most influences whether or not you accept the above argument?

A No reasonable mechanism exists which would allow all the original helium and hydrogen to be lost from the atmosphere.

B The Earth's gravitational field is not strong enough to hold the light gases helium and hydrogen and they eventually diffuse away into space.

C Hydrogen is a lighter element than helium.

D If the argument is correct there should be no hydrogen in the atmosphere.

E Helium is the first element in group 18.

8 | The 'inert pair effect' describes the observation that the *s* electrons have a decreasing tendency to become involved in bonding as a group is descended within the *p*-block of the periodic table. This is illustrated by the following observations: Ge(II) is a strong reducing agent whereas

Ge(IV) is stable; Sn(II) exists as reducing ions whereas Sn(IV) is covalent and stable; Pb(II) is much more stable than Pb(IV).

Which one of the following statements best summarises the general conclusion that can be drawn from this passage?

A The +2 oxidation state becomes increasingly stable as group 14 is descended.

B The inert pair effect is due to the inherent 'inertness' of the two *s* electrons.

C Lead(II) compounds are ionic.

D Lead(IV) is oxidising.

E Carbon and silicon are stable in oxidation state (IV) only.

9 One way of reducing the levels of CO, NO_x and hydrocarbons in the exhaust gases from car engines is to fit catalytic converters. The most effective catalyst is made of platinum, and this complexes with lead which therefore poisons the catalyst. Lead-free petrol was therefore developed in order to make the use of catalysts possible.

Which one of the following sentences best expresses the flaw in this argument?

A It is possible to remove lead from platinum.

B Lead is present in engine oil, and so this will poison the catalyst.

C Nowadays the best catalysts contain rhodium as well as platinum.

D Lead-free petrol was developed because of the potential health hazard of lead emissions from the exhaust.

10 Which is more soluble in water, $NaClO_4$ or $KClO_4$? The most important concept to remember is the general rule that compounds containing ions with widely different radii are more soluble in water than compounds containing ions with similar radii. The six coordinate radii of Na^+ and K^+ are 0.102 and 0.138 nm respectively. Therefore, the salt $NaClO_4$ should be more soluble in water than $KClO_4$.

Which one of the following is an underlying assumption in the argument?

A Water is a polar solvent.

B Potassium has a higher atomic number than sodium.

C The radius of the perchlorate ion is greater than 0.14 nm.

D Na^+ and K^+ are in the same group in the periodic table.

E Na^+ is more electronegative than K^+.

11 In graphite, only three of the valence electrons are involved in forming σ bonds using sp^2 hybrid orbitals. This gives rise to a sheet structure within which the remaining electron forms a π bond. A single crystal of graphite conducts electricity in the two dimensions which form the sheet, but not in the third. Therefore conduction of electricity occurs within a sheet but not from one sheet to another.

Which one of the following statements best represents a theoretical explanation for the conclusion drawn in this passage?

A Delocalisation of π electrons in graphite takes place in two dimensions rather than in three dimensions.

B The distance between sheets in graphite is too large for electrons to 'jump'.

C π bonds must be present for conduction of electricity.

D sp^2 hybridisation gives rise to sheet structures.

12 You cannot take a logarithm of a unit. Therefore, in the equation

$$\Delta G^{\ominus} = -RT \ln K$$

$\ln K$ has no units. For a reaction

$$A \rightleftharpoons B + C$$

the equilibrium constant has units of mol dm^{-3}. Therefore the equation for $\Delta G^{\ominus}$ can only apply to reactions in which the number of reactants and products are equal.

Which one of the following is the best statement of the flaw in this argument?

A Units are always ignored when logarithms are taken.

B The units of K depend on the relative numbers of reactants and products in a reaction.

C When calculating an equilibrium constant it is necessary to use activities and not concentrations.

D There is no flaw in this argument.

13 Two analysts each use the same procedure to make five replicate measurements of the percentage of carbon in the same sample. The mean values obtained by the two analysts were 51.2% and 50.9% respectively.

However, the standard deviation of Analyst A's readings was 3.7, and for Analyst B's readings it was 4.5. Therefore Analyst A is the more consistent worker.

Which one of the following is the best statement of the flaw in the argument?

A There may be a systematic error in the measurement.

B The difference between these two standard deviations is not big enough for the two values to be significantly different from each other.

C The standard deviation gives no information about accuracy.

D The confidence limits of the mean are given by the standard error of the mean and not by the standard deviation.

14 Salts of carboxylic acids react with cyanogen bromide to give alkyl nitriles according to the following equation:

$$RCOONa + BrCN \rightarrow RCN + CO_2 + NaBr$$

A proposed mechanism involves the attack of the cyanide ion on the carboxylate salt with displacement of the COO^- group.

Which one of the following statements most weakens this proposal?

A If the starting material has ^{14}C in the carboxylate group, the resulting alkyl nitrile is labelled with ^{14}C.

B Since the carboxylate salt has changed into CO_2 and Na^+, the mechanism must be more complicated than the one proposed.

C The yield is not 100% so other reactions must be occurring.

D The reaction works better when R is aromatic than when R is aliphatic.

15 During the work-up of the products from the hydrolysis of a phenyl ester an aqueous solution is obtained. This is expected to contain an acid, a phenol and an involatile neutral substance. In order to separate the constituents, ether and dilute alkali are added and the mixture is then shaken. The layers are separated and the ether is evaporated to dryness but no residue is present. It is concluded that there is no neutral substance present in the products being worked up.

Which one of the following is the best statement of an assumption underlying this conclusion?

A That the phenol will not esterify in aqueous alkali.

B That the acid does not form an insoluble sodium salt.

C That the neutral substance is more soluble in ether than in the aqueous phase.

D That the phenol is insoluble in the ether.

E That water and ether are substantially immiscible.

F That the acid is insoluble in ether.

16 Most compounds of transition metals are coloured. This can be explained by the fact that atoms and ions of transition metals typically have partially filled d orbitals so that d–d transitions are possible. No such transitions are possible in compounds such as AgCl and TiO_2 since the d orbitals are respectively completely filled and empty. These compounds are white. This confirms that the origin of colour in compounds of transition metals is the d–d transition.

Which one of the following statements is the best expression of the flaw in this argument?

A AgBr and AgI are yellow.

B Polarisation of large anions can lead to covalency and coloration.

C $KMnO_4$ and K_2CrO_4 are coloured.

D Intense colours are produced by charge transfer processes.

E Nonstoichiometric oxides of transition metals are black.

F d–d transitions may not be the only electronic transitions taking place.

17 The structure of ethyl 3-oxobutanoate can be represented as a ketone or as an enol:

$$CH_3COCH_2CO_2C_2H_5 \quad \text{or} \quad CH_3C(OH)=CHCO_2C_2H_5$$

The correct form can be deduced to be the enol, since it gives a red colour with Fe^{3+}, and this is characteristic of enols.

Which one of the following is the best statement of the flaw in this argument?

A No information is given about the way in which the test was carried out.

B Many compounds give a red colour with Fe^{3+}.

C The test should also have been carried out on control substances which are known to contain an enol and known not to contain an enol.

D Compounds capable of exhibiting keto-enol tautomerism will almost always react as enols with Fe^{3+}, whatever the equilibrium position of the tautomerisation.

E Enols have lots of other characteristics besides giving a red colour with Fe^{3+}.

18 The Kelvin scale of temperature defines a degree as 0.01 of the temperature difference between the freezing point and the boiling point of water at one atmosphere of pressure. On the Fahrenheit scale, water freezes at 32 °F and boils at 212 °F so that the rise in temperature for 1 °F on the Kelvin scale

I PREFER THE KELVIN SCALE
... SEEMS WARMER ...

is $100/(212-32)$ or 5/9 of the rise for one degree on the Kelvin scale. Absolute zero on the Fahrenheit scale is $(32 - 9/5 \times 273)$ or −459.4 °F. This scale could be converted into an absolute scale (A) with 0 at absolute zero; on this scale, water would freeze at $(459.4 + 32)$ or 491.4 A and would boil at $(459.4 + 212)$ or 671.4 A. The equation $\Delta G^{\ominus} = -RT \ln K$ is a fundamental equation which must hold for any temperature scale. It follows that, using the absolute scale (A) the gas constant (R) would have the value 14.965 J mole^{-1} A^{-1}.

Which one of the following would you do to convince yourself of this argument?

A Check that the formula F = 32 + (C × 9/5) gives a value of −459.4 for absolute zero defined as −273 °C.

B Check the definition of $\Delta G^{\ominus}$.

C Check that the gas constant for the Kelvin scale is $8.314 \, J \, K^{-1} \, mol^{-1}$ and that $8.314 \times 9/5$ is 14.965.

D Find out how the zero point on the Fahrenheit scale was originally defined.

19 Some support materials used to make ion exchange columns carry sulfonic acid groups (RSO_3^-). These will therefore form ionic bonds with any substance which carries a positive charge. The strength of the interaction depends on the average value of the positive charge on each molecule. Two compounds x and y are both quaternary ammonium salts and carry no other ionisable group. It follows that both will bind equally well to a cation exchanger such as those carrying sulfonic acid groups, and that they cannot be separated using that medium.

Which one of the following statements best represents the flaw in the argument?

A Ion exchange resins do not interact with solute molecules solely through ionic interactions.

B Different ion exchange resins carry different ionisable groups.

C The two substances might be separated with a column containing a non-ionic solid phase (such as alumina or cellulose).

D Adjusting the pH may result in a change in the charge on ionisable groups and this may affect one of the compounds x or y more than the other.

20 Chlorine atoms, formed in the stratosphere by the photochemical degradation of chlorofluorocarbons, react with the ozone which is present to give O_2 and ClO. ClO reacts further with oxygen atoms to give O_2 and to regenerate chlorine atoms. In this way, a single chlorine atom can cause the destruction of a large number of ozone molecules.

Which one of the following is the best statement of the underlying assumption in this argument?

A The chain reaction by which chlorine atoms are regenerated is terminated when a chlorine atom reacts with some species other than ozone found in the stratosphere.

B The reaction between chlorine atoms and ozone is first order with respect to both reactants.

C The ratio of O_3:Cl in the stratosphere is very large.

D There is a source of oxygen atoms in the stratosphere.

21 The properties of benzene cannot be accounted for by any chain structure with an empirical formula C_6H_6. In 1865 Kekulé proposed that benzene is a six-membered ring with alternating single and double bonds. This accounted for the fact that each carbon atom in benzene was found to be equivalent, but did not explain why benzene does not react like an alkene. The explanation is that the electrons in the ring cannot be localised in particular double and single bonds, but must be delocalised (i.e. shared equally between all six carbons).

Which one of the following, all of which are true, most strengthens the conclusion of the argument?

A Bromine adds across the double bond of an alkene, but reacts with benzene to give the substitution product, bromobenzene.

B X-ray diffraction studies on crystals of benzene show that all the C–C bonds are of equal length.

C NMR spectroscopy shows that the protons are all equivalent and the signal indicates that they are extremely deshielded.

D The molecular orbital theory of bonding was developed partly as a result of the need to explain the behaviour of electrons in aromatic compounds.

22 Two different products may be obtained when naphthalene is sulfonated. At room temperature, the product is naphthalene-1-sulfonic acid, whereas at 165 °C the product is naphthalene-2-sulfonic acid. Heating naphthalene-1-sulfonic acid with sulfuric acid at 180 °C converts it into naphthalene-2-sulfonic acid.

Which one of the following statements is the most sensible logical inference to draw from the above observations?

A Sulfonation of aromatic compounds is a good way of obtaining a variety of products.

B At room temperature, the sulfonation of naphthalene is kinetically controlled.

C Electrophilic aromatic substitutions often give mixtures of products.

D The formation of products in aromatic substitutions is very sensitive to the experimental conditions.

E The relative thermodynamic stabilities of the two products are dramatically affected by temperature.

23 You have obtained a set of ten x, y values and tabulated them in decreasing order of x values. Inspection of the table shows that y values are also listed in decreasing order. You calculate the correlation coefficient (R). The value obtained is 0.98, compared with a value of 1.00 which would be obtained for data which fit a perfect straight line with positive slope.

In order to convince yourself that x and y are linearly correlated, which of the following would you do?

A Obtain more x, y values and recalculate the correlation coefficient.
B Use a statistical test to check whether, in this context, the R value of 0.98 is significantly less than 1.
C Plot the data as a graph.
D Calculate values for the slope, the intercept and their 95% confidence limits.

24 Water dissociates according to the equation:

$$H_2O \rightleftharpoons H^+ + OH^-$$

pH is defined as $-\log_{10}[H^+]$, and neutral pH as that at which $[H^+]$ and $[OH^-]$ (the activities of H^+ and OH^-) are equal. Under standard conditions, the product of the activities is found to be 10^{-14}. It follows that neutral pH is defined as 7.0.

What is the best statement of the flaw in this argument?

A The product of $[H^+]$ and $[OH^-]$ is temperature dependent and therefore neutral pH is only 7.0 at 25 °C.
B You can take a log of a number, but not of a quantity with dimensions.
C Water dissociates to give OH^- and H_3O^+.
D Pure water has a molar concentration of about 55.5 mol dm^{-3} and this should be taken into account in determining the equilibrium constant for the dissociation of water (55.5=1000/18).

25 A potential customer requires an assurance from the supplier of a chemical product that the product contains less than 20 ppm of a known by-product of the synthesis. No other quality standards are specified. The product is analysed in the supplier's laboratory and the level of the by-product is quoted as 18.4 ± 1.5 ppm.

Which one of the following conclusions can you draw from this information?

A The known by-product must be extremely hazardous.

B The analytical method is not good enough to determine whether the product meets the customer's specification.

C The supplier can confidently give the customer the required assurance.

D The supplier would be running a high risk in giving the customer the required assurance.

E There is not enough information to decide whether or not the supplier should give the required assurance.

26 Sulfur dioxide reacts with water droplets in the atmosphere to form H_2SO_3. This will dissociate in accordance with the equation:

$$H_2SO_3(aq) + H_2O(l) \rightleftharpoons H_3O^+(aq) + HSO_3^-(aq)$$

The dissociation constant (K_a) for this reaction has a value of 10^{-2}. If sufficient sulfur dioxide is dissolved so that the ratio of H_2SO_3:HSO_3^- is exactly 1, then the concentration of H_3O^+ will be $K_a \times [H_2O(l)]$. The concentration of water in dilute aqueous solution is about 55 mol dm^{-3} so that $[H_3O^+]$ is 5.5×10^{-1} or approximately $10^{-0.4}$. Since $pH = -\log_{10}[H_3O^+]$, the pH of the solution will be about 0.4.

Which one of the following sentences is the best statement of the flaw in this argument?

A The concentration of water is defined as 1.

B The ratio of H_2SO_3:HSO_3^- must be constant and cannot be affected by the amount of dissolved SO_2.

C You can never be certain that the ratio of H_2SO_3:HSO_3^- is exactly equal to 1.

D K_a is a dissociation constant, not an equilibrium constant, and the concentration of water is therefore subsumed in it.

27 The equation $\Delta G = \Delta H - T\Delta S$ implies that a process for which ΔS is positive will become more unfavourable as temperature increases. Pentanol is only sparingly soluble in water because the alkane chain cannot form hydrogen bonds with solvent water. Nevertheless the total number of hydrogen bonds between water molecules is not much affected by adding pentanol in solution. This is because the water molecules rearrange themselves into an ordered cage round the alkane chain. The decrease in entropy caused by this ordering explains the limited solubility of molecules like pentanol.

Which one of the following statements most strengthens the above argument?

A Solubility generally tends to increase with temperature.

B Pentanol is more soluble in cold water than in hot.

C The number of hydrogen bonds between water molecules decreases with temperature.

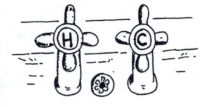

D Disorder is generally increased when two pure compounds are mixed.

E A positive value of ΔS represents a decrease in order of the system.

28 The reaction between OH radicals and NO_2 plays an important role in the chemistry of the stratosphere. Pseudo first order rate constants (k_{1st}) for this reaction have been determined in the laboratory from the rate of disappearance of OH radicals in the presence of excess levels of NO_2. N_2 was used to maintain pressures found in the stratosphere. In these experiments, the sum of the partial pressures of OH and NO_2 made up less than 0.1% of the total pressure of the system. Second order rate constants were obtained from plots of k_{1st} against the partial pressure of NO_2. The second order rate constant was found to increase with pressure.

Which one of the following sentences best expresses the conclusion you would draw from this statement?

A The concentration of OH radicals in the stratosphere is very low.

B The formation of HNO_3 is a two-step process involving collision with another molecule, such as N_2.

C The study of reaction kinetics can always give useful information about reaction mechanisms.

D It is important to avoid polluting the atmosphere with oxides of nitrogen.

29 Benzene is reduced to cyclohexa-1,4-diene by treatment with sodium in liquid ammonia in the presence of ethanol and HCl. The reaction is thought to proceed according to the following mechanism shown below.

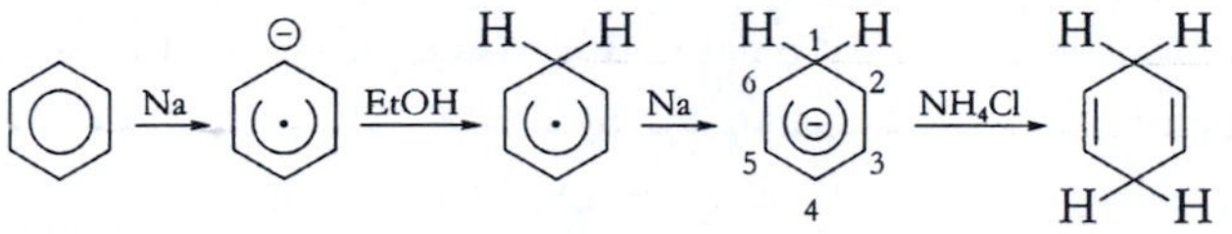

(I)   (II)   (III)   (IV)   (V)

On thermodynamic grounds the conjugated product cyclohexa-1,3-diene (VI) is expected to be more stable than the cyclohexa-1,4-diene (V). Thus the fourth step involving the electrophilic attack by NH_4^+ must be kinetically controlled. It follows that the site of highest electron density in IV is on carbon 4.

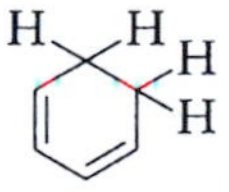

(VI)

Which one of the following studies would you carry out in order to confirm this conclusion?

A Thermochemical studies on cyclohexa-1,3-diene and on cyclohexa-1, 4-diene.

B Determination of the rate of the reaction.

C ^{13}C NMR study of the anionic intermediate.

D Base catalysed equilibrium studies on cyclohexa-1,3-diene and cyclohexa-1,4-diene.

30 Two molecules of ethanol can hydrogen bond to each other via the lone pair of electrons on the oxygen of one and the hydroxyl hydrogen of another. The strength of this type of hydrogen bond is about 20 kJ mol^{-1}.

Thus the equilibrium constant (K) for the formation of a hydrogen bond is equal to $e^{20/RT}$ (or about 1000 at room temperature). This suggests that in pure ethanol about 99.9% of the molecules will be hydrogen bonded to other ethanol molecules.

Which one of the following is the best statement of the flaw in this argument?

A The strength of a hydrogen bond is a measure of $\Delta H^\ominus$ not $\Delta G^\ominus$.

B The equation $\Delta G^\ominus = -RT \ln K$ does not apply to pure liquids.

C In a pure liquid the concentration is defined as 1.

D The true value of the equilibrium constant is $e^{-20/RT}$ or about 1/1000.

E There is no flaw in this argument.

CONSTRUCTING
AN ARGUMENT

An argument is a sequence of ideas (sentences) which lead to a logical conclusion. The construction of arguments is part of the process by which experimental observations are interpreted, linked with existing knowledge, and communicated to others. Each exercise in this section involves the construction of an argument by arranging three sentences in a logical sequence and adding suitable linking words between the second and third sentence. Not all the examples lead to an argument with the same structure. Possible structures include:

* X, Y, therefore Z.
* X, Y, this illustrates or demonstrates Z.
* X, Y, this explains why Z.

A complete solution to each exercise involves

* Writing down in a logical sequence (the letters representing) each sentence.
* Writing down appropriate linking words between the second and third sentence.
* Rehearsing the reasons for preferring the chosen sequence to any other.

In deciding on your preferred solution, bear in mind two points.

First, in a well-written passage, the choice of words used in a sentence depends on the position of the sentence in the argument. For example, a passage about a particular compound will refer to it by name in the first sentence, and probably as 'it' in the second sentence. In these passages the wording has been changed in order to avoid this kind of clue being a strong influence. This does not mean that there are no linguistic clues in the sentences, since comparatively small differences in sentence structure can have an important effect on their meaning.

The second point is that the restriction of having no linking words between the first and second sentences is not a rule of argument or of grammar. We have limited the structure of argument to this form simply to prevent preferred sequences becoming obscured by too many alternative formulations of the kind

A. This is illustrated by B. This supports the hypothesis C.

Imaginative use of this type of structure may make it possible to create several logical sequences from the sentences in these exercises. It is therefore a useful discipline to allow linking words only between the second and third sentences.

In order to get started, we have provided a commentary on the first exercise. In Section 6 you will find commentaries on all the rest of the odd-numbered exercises.

1 **A** When phenol is treated with bromine water, the product is 2,4,6-tribromophenol.
  B Phenol reacts with electrophiles.
  C The OH group is *ortho-para* directing.

GETTING STARTED

Our preferred sequence is B, A, therefore C.

Phenol reacts with electrophiles. When phenol is treated with bromine water, the product is 2,4,6-tribromophenol. Therefore the OH group is ortho-para directing.

The sequence 'B, C, therefore A' may seem attractive because it has a perfectly logical structure. However, the logic is false: sentences B and C together lead to the prediction that the product of the reaction would be a mixture of different ortho and para derivatives, but this is not the observation given in A.

An alternative to 'B, A, therefore C' is 'A, B, therefore C'. In our view, the former reads more easily; it seems natural to start with the generalisation (B), and to follow this with a particular example (A), from which the conclusion (C) is drawn.

Several linking phrases are possible; these include 'this shows that' or 'this is evidence that'.

2 **A** The synthesis of fertiliser grade (i.e. impure) phosphoric acid is cheaper than the synthesis of pure phosphoric acid.
  B The synthesis of pure phosphoric acid involves several synthetic steps including the sublimation of white phosphorus, P_4.
  C Fertiliser grade phosphoric acid can be synthesised in a single step because it does not need to be of high purity.

3 **A** Compounds containing a single chiral carbon atom have two enantiomers each of which rotates plane-polarised light.

B A carbon atom is described as chiral when it is covalently bonded to four different substituents.

C Each enantiomer of lactic acid rotates plane-polarised light.

$$H-\underset{\underset{OH}{|}}{\overset{\overset{COOH}{|}}{C}}-CH_3 \qquad \text{Lactic acid}$$

4 **A** Lactic acid has two enantiomers and is described as a chiral molecule.

B Naturally occurring lactic acid consists of one of two possible enantiomers, and it rotates plane-polarised light.

C Chiral molecules with a single asymmetric centre rotate plane-polarised light.

5 **A** For mononuclear oxoacids, the species with the greater number of oxo groups has the lower pK_a.

B $HClO_4$ is a stronger acid than $HClO_3$.

C $HClO_4$ has more oxo groups than $HClO_3$.

6 **A** Zinc hydroxide is amphoteric.

B Addition of $NaOH(aq)$ to $ZnCl_2(aq)$ gives a white precipitate which redissolves in excess alkali.

C Zinc hydroxide is soluble in dilute hydrochloric acid.

7 **A** Graphite has a layer structure.

B Graphite is soft and can be used as a lubricant.

C The layers of graphite are held together by van der Waals forces.

8 **A** A methyl group in a benzene ring is relatively electron-donating (compared with H).

B Nitrating agents such as the nitronium ion (NO_2^+) attack electron-rich sites.

C The rate of nitration of methylbenzene is greater than the rate of nitration of benzene.

9 **A** A solution containing thiocyanate gives a red colour with the Fe^{3+} ion.

   B Solution X gives a red colour with Fe^{3+}.

   C Solution X contains thiocyanate.

10 **A** Platinum is used as a catalyst in catalytic converters in petrol engines.

   B Lead has a strong affinity for platinum and deactivates its catalytic activity.

   C Manufacturers specify that cars fitted with catalytic converters must run on lead-free petrol.

11 **A** Methanol forms hydrogen bonds and is a liquid.

   B The ability to form hydrogen bonds may have an important effect on the physical properties of a compound.

   C Methane does not form hydrogen bonds and is a gas.

12 **A** The hydrolysis of adenosine triphosphate to adenosine diphosphate and phosphate ions in neutral aqueous solution at room temperature occurs rapidly in the presence of an appropriate enzyme.

   B Adenosine triphosphate is thermodynamically unstable but kinetically stable in neutral aqueous solution at room temperature.

   C An aqueous solution of pure adenosine triphosphate is stable at room temperature.

13 **A** The equilibrium constant for a reaction is defined as the product of the activities of products divided by the product of the activities of reactants, when the system is at equilibrium.

   B For the reaction A $\rightleftharpoons$ B, the equilibrium constant is 1.

   C When A and B are in equilibrium, their activities are equal.

14 **A** The C–N bond length in amides, —C(O)—NH—, is less than that in saturated amines, —CH_2—NH_2, and greater than that in imines, —CH=NH.

   B An amide is best represented as a resonance hybrid.

   C A resonance hybrid structure is a way of representing some molecules for which more than one valence structure can be drawn, and which have properties that are not adequately represented by any single valence structure.

15

A Ammonium salts of nitrate, chlorate and perchlorate are unstable and readily decompose when heated.

B A common feature of the nitrate, chlorate and perchlorate ions is that they are strong oxidants.

C Ammonium salts of oxidising anions are unstable.

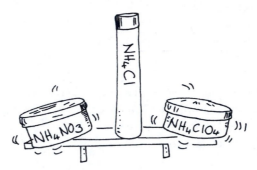

16

A When silver chlorate decomposes there is an increase in entropy.

B Gases are more disordered than crystalline solids.

C Silver chlorate, which is a crystalline solid, decomposes according to the following equation:

$$2AgClO_3(s) \rightarrow 2AgO(s) + 2O_2(g) + Cl_2(g)$$

17

A The sum of the covalent radii for P and O is 0.176 nm.

B In phosphorus pentoxide, P_2O_5, different P–O bonds have bond lengths of 0.160 and 0.143 nm respectively.

C P and O are capable of forming a $d\pi$–$p\pi$ bond.

18

A The efficiency of utilisation of visible solar radiation for photosynthesis is limited by factors such as the quantum efficiency of photosynthesis, reflection of light from leaves, photorespiration, and lack of leaf cover at certain times of the year.

B Only a small fraction of the solar radiation reaching the ground can be recovered as chemical energy in crops.

C The energy needed to support photosynthesis is supplied by the electromagnetic radiation in the visible region of the spectrum, and this makes up about half of the total solar radiation incident on the Earth's surface.

19
A Many Co(III) complexes are inert to substitution.
B Many Co(III) complexes are octahedral and have low spin d^6 electron configurations.
C Octahedral complexes with low spin d^6 electron configurations are inert to ligand substitution.

20
A Compounds of d block elements are usually coloured because of electronic transitions between d orbitals.
B Zn^{2+} compounds are white.
C Zn^{2+} compounds have a completely filled set of d orbitals.

21
A Saturated aliphatic esters absorb in the infrared region at around 1740 cm^{-1}.
B Compound A absorbs infrared radiation at around 1740 cm^{-1}.
C Compound A is a saturated aliphatic ester.

22
A ^{13}C nuclei can be detected directly by their nuclear magnetic resonance spectrum, but ^{12}C nuclei cannot.
B Reactants in which selected carbons are enriched in ^{13}C are useful for investigating reaction mechanisms.
C ^{13}C is a stable isotope of carbon possessing virtually the same chemical properties as the abundant isotope ^{12}C.

23
A $SiCl_4$ is a tetrahedral molecule.
B $SiCl_4$ is sp^3 hybridised.
C sp^3 hybridisation gives rise to tetrahedral molecules.

24
A O_2 is paramagnetic.
B Paramagnetism is caused by unpaired electrons.
C O_2 has unpaired electrons.

AND I'D LIKE TO BE BURIED INSIDE A LARGE MODEL OF THE SILICON TETRACHLORIDE MOLECULE …

25 **A** Only compounds with an asymmetric electron distribution have a dipole moment.

   B $Ge(CH_3)_4$ does not have a dipole moment.

   C $Ge(CH_3)_4$ is spherically symmetric.

26 **A** Titanium(IV) has no valence electrons, but titanium(III) has one, so that the substitution of a Ti^{3+} ion for a Ti^{4+} ion in a lattice adds an electron to the lattice.

   B An n-type semiconductor is created by introducing an excess of negatively charged electrons into a lattice.

   C A Ti^{4+} lattice doped with Ti^{3+} ions is an n-type semiconductor.

ACTUALLY THEY'RE NEGATIVELY CHARGED LETTUCES

27 **A** BF_3 readily forms an adduct with ammonia.

   B BF_3 has six valence electrons.

   C Ammonia is a Lewis base.

28 **A** Under standard conditions titanium(IV) cannot be reduced to titanium(III) by SO_3^{2-}.

   B Under standard conditions vanadium(V) can be reduced to vanadium(IV) by SO_3^{2-}.

   C The vanadium(V)/vanadium(IV) couple has a higher standard reduction potential than titanium(IV)/titanium(III).

29 **A** Salts containing anions and cations of widely different radii are generally more water soluble than salts containing ions with similar radii.

B The (six coordinate) radii of Na^+ and K^+ are 0.102 nm and 0.138 nm respectively.

C Potassium salts of large anions are generally less soluble than the corresponding sodium salts.

30 **A** After treatment with D_2O, amides do not have an absorbance band in the region of 3200 cm^{-1}.

B The N–H bond in amides exchanges readily with hydrogen from water.

C Amides have an infrared absorbance peak in the region of 3200 cm^{-1} which is attributed to the stretching of the N–H bond.

31 **A** The Haber process used to manufacture ammonia from nitrogen and hydrogen is carried out at a temperature between 400 and 500 °C.

B The reaction

$$3H_2 + N_2 \rightleftharpoons 2NH_3$$

is exothermic, so le Chatelier's principle indicates that ammonia production would be favoured by low temperatures.

C At room temperature, the rate of reaction between nitrogen and hydrogen is too slow to be commercially useful.

32 **A** The solubility of $MgCl_2$ in water is greater than that of $BaCl_2$.

B Mg^{2+} is smaller than Ba^{2+} and therefore it has a higher charge density.

C Water molecules interact more strongly with Mg^{2+} than with Ba^{2+}.

33 **A** Many molecules with several conjugated double bonds are coloured.

B The energy required to excite an electron in molecules with conjugated double bonds is often between 2×10^{-19} and 5×10^{-19} Joules.

C The energy of a photon from the visible region of the spectrum lies between 2×10^{-19} and 5×10^{-19} Joules.

34 **A** The activity of a substance in solution is defined as:

concentration × activity coefficient / the standard state

B An equilibrium constant should be determined from activities and not from concentrations.

C An equilibrium constant has no dimensions.

35 **A** The density of graphite is lower than the density of diamond.

B In graphite, two-dimensional sheets of carbon atoms are separated by a distance of approximately 0.335 nm.

C Each carbon atom in a two-dimensional sheet of graphite is bonded to three others in the plane (bond length 0.140 nm) whereas in diamond the carbon atoms are tetrahedrally arranged and the bond length is 0.154 nm.

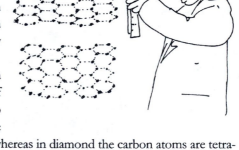

36 **A** An allyl carbocation can be described by more than one resonance structure.

B An allyl carbocation is more stable than a propyl carbocation.

C Resonance stabilisation plays an important part in the stability of carbocations.

37 **A** H_3AsO_4 reacts with iodide in aqueous solution to form an equilibrium mixture with $HAsO_2$ and iodine.

B When excess tin(II) chloride is added to H_3AsO_4 in the presence of iodide all of the H_3AsO_4 is converted into $HAsO_2$.

C Tin(II) chloride reduces iodine to iodide.

38 **A** When carried out at constant temperature and pressure, the reaction between H_2S and oxygen has a favourable enthalpy term.

B When H_2S and oxygen react according to the equation given below, $\Delta G^{\ominus}$ is negative.

$$2H_2S(g) + 3O_2(g) \rightarrow 2H_2O(l) + 2SO_2(g)$$

C Entropy decreases when the number of molecules decreases during a reaction and when liquids are produced from gases.

39 **A** The enthalpy of formation of $C_8H_{18}(g)$ is -169 kJ mol^{-1}.

B The enthalpy of formation of $H_2O(g)$ is -242 kJ mol^{-1} and that for $CO(g)$ is -111 kJ mol^{-1}.

C The value of $\Delta H^{\ominus}$ for the formation of octane (C_8H_{18}) from gaseous H_2 and CO is -1217 kJ mol^{-1}.

$$8CO(g) + 17H_2(g) \rightarrow C_8H_{18}(g) + 8H_2O(g)$$

40 **A** α-particles are positively charged and those from natural sources are slow moving.

B Elements of large atomic number have nuclei with a large positive charge which repel slow-moving α-particles.

C Only α-particles which have been accelerated bring about nuclear re-actions in elements with a large atomic number.

ALPHAS ARE SLOW BUT POSITIVE

α

CRITICAL READING

All authors must make assumptions about their readers, and all readers are different. Consequently it is always possible that the reader's needs are not met exactly by the writer's words, and so the reader must always be prepared to interpret the written word. Interpretation is especially important when one is reading in order to learn rather than reading for pleasure.

Interpretation often means asking questions about the written word.

* Has the writer made an assumption that you know some fact or understand some principle which you actually do not?
* Have you appreciated how new ideas integrate with what you already know?
* Has the writer simplified something more than you think is reasonable?
* Has the writer missed out some important evidence supporting an idea being introduced; do you want to ask 'how can that be known'?
* Do you think the author is wrong? (This is always possible; but it is also possible that you yourself have some muddled ideas which make what is written *seem* wrong; so it is always worth questioning whether what is written *seems* right.)

It takes experience and skill to recognise whether or not one has fully understood a particular passage. It takes courage and self-confidence to question what one has read. The exercises in this section are designed to help you to gain the necessary skills and confidence to develop into a critical (and therefore effective) reader.

Notice that 'critical' does not mean adverse criticism of the writer or the text. It is true that mistakes can occasionally be found in all types of scientific literature – including undergraduate textbooks and laboratory manuals. But we are not concerned with these. Indeed we have rejected examples of writing which in our view are incorrect, misleading, ambiguous, or written in poor style. 'Critical' in this section means reading alertly.

Each exercise is based on a passage taken from a book, magazine or journal (sometimes carefully abridged). In some cases the task is to select one response from those given. In other cases specific questions are asked, to which simple answers can be provided in a few words. Some of the exercises (27–37) involve interpretation of numerical data. Other exercises include the task of writing a paragraph which explains the passage for a person without a scientific background; this is not just a case of paraphrasing the passage, but of interpreting it in an appropriate way. The end of the section includes a few exercises which involve writing for other scientists; these examples provide descriptions of experimental methods given in papers which should give 'sufficient information for someone else to repeat their work'. You will discover that, in order to expand these descriptions into a set of instructions, you will have to use some creative thinking.

As with the previous two sections we have included a commentary on the first exercise in order to help you to get started. However, because of the different style of exercise in this section, the commentary on Exercise 1 is unlikely to be of direct help in dealing with all the exercises. If you find the later exercises more difficult, you need to refer to the commentaries at the end of the book.

1 Until recently, carbon was thought to occur in only two principal allotropic forms: diamond and graphite. Both allotropes are covalent network solids, whose structures we discussed in detail previously. In diamond, each carbon atom is tetrahedrally (sp^3) bonded to four other carbon atoms. To move one plane of atoms in the diamond crystal relative to another requires the breaking of many strong carbon–carbon bonds. Because of this, diamond is one of the hardest substances known. As a pure substance, diamond is colourless, although natural diamond may be coloured by impurities.

D D Ebbing, *General Chemistry* (5th edition)
Houghton Mifflin, 1996

Which one of the following statements best represents an underlying assumption in this passage?

A In diamond the interatomic bonds are more numerous per unit volume and/or stronger than interatomic bonds in almost all other substances.

B Diamond has a hardness of 10 on Mohs' scale.

C New allotropes of carbon (the fullerenes) have now been discovered.

D Coloured diamonds are less hard than colourless diamonds.

GETTING STARTED

Our choice is A.

The passage says that moving a plane of atoms in diamond involves breaking many 'strong carbon–carbon bonds'. However, a similar statement could be made about other giant lattice structures, for example silicon ($SiO_2)_n$. The implication of diamond being 'one of the hardest substances known' is that it is more difficult to break the carbon–carbon bonds than it is to break bonds in other giant lattice structures. The carbon–carbon bonds must therefore either be stronger or more frequently encountered (more densely distributed) than bonds in other giant lattice structures. The author of this passage is assuming that this is recognised.

B is true, but is not an underlying assumption of the passage; it is a semi-quantitative statement of the hardness of diamonds.

C is true, and it is implied by the first sentence of the passage. This does not make it an underlying assumption.

D is an interesting point since, as the passage states, coloured diamonds contain impurities and these must create defects in the lattice structure. Since the strength of a substance is limited by its weakest links, coloured diamonds must be some-what less hard than colourless ones. However, our view is that the final sentence in the passage is not related to the main argument about the hard-ness of diamond, and so D is not an underlying assump-tion, even though it must tech-nically be true. (It does not follow that the impurities in coloured diamonds are suffi-cient to cause significant loss of hardness.)

2 Scandium is perhaps as similar to aluminium as to yttrium and the lanthanides because of its small ionic radius (0.68Å) . . . The fluoride is insoluble in water but dissolves readily in an excess of HF or in NH_4F to give fluoro complexes such as $[ScF_6]^{3-}$, and the similarity to Al is confirmed by the existence of a cryolite phase Na_3ScF_6.

F A Cotton & G Wilkinson, *Advanced Inorganic Chemistry* (5th edition)
John Wiley & Sons, 1988

Which one of the following statements is an underlying assumption in this passage?

A Scandium and aluminium have similar radii.

B Aluminium occurs naturally as cryolite.

C Cryolite is Na_3AlF_6.

D Cryolite is insoluble in water but dissolves in HF.

E The cryolite structure is adopted by many salts containing small cations and large anions.

3 Tollen's test for distinguishing between aldehydes and ketones relies on the difference in ease of oxidation of these two classes of compounds. In this test, silver ion in ammonia solution functions as a mild oxidising agent. Silver(I) is reduced to silver metal by aldehydes but not by ketones. To test a compound, you put a sample and Tollen's reagent in a clean test tube or flask. If the compound is an aldehyde, silver metal will deposit in a few minutes on the inside of the glass vessel as a reflective coating.

D D Ebbing, *General Chemistry* (5th edition)
Houghton Mifflin, 1996

Which of the following statements are plausible reasons for using ammonia solution?

A It makes the solution alkaline.
B It complexes the silver ion.
C It acts as a mild detergent to clean the glass surface
D It neutralises the acid formed by oxidation of the aldehyde.
E It complexes the aldehyde.
F It protects the silver film from redissolution as it forms.
G It acts as a secondary reducing agent.

4 In the vapour phase, alkali metal halides are present mainly as ion pairs, but measurements of bond lengths and dipole moments suggest that under these conditions electron-sharing is also involved to a considerable extent, especially for the lithium halides.

A G Sharpe, *Inorganic Chemistry* (3rd edition)
Longman, 1992

Which one of the following statements best represents the key point of passage?

A Alkali metal halides are predominantly ionic.
B Measurement of bond lengths and dipole moments are effective methods for estimating electron-sharing.
C Alkali halides exhibit some covalent character.

5 A molecule within the body of a liquid tends to be attracted equally in all directions, so that it experiences no net force. On the other hand, a molecule at the surface of a liquid experiences a net attraction by other molecules toward the interior of the liquid. As a result, there is a tendency for the surface area of a liquid to be reduced as much as possible. This explains why falling rain-drops are nearly spherical.

Energy is needed to reverse the tendency toward reduction of surface area in liquids. Surface tension is the energy required to increase the surface area of a liquid by a unit amount.

The surface tension of a liquid can be affected by dissolved substances. Soaps and detergents, in particular, drastically decrease the surface tension of water. An interesting but simple experiment shows the effect of soap on surface tension. Because of surface tension, a liquid behaves as though it had a skin. You can actually float a pin on water, if you carefully lay it across the surface. If you then put a drop of soap solution onto the water, the soap spreads across the surface and the pin sinks. Water bugs also sink in soapy water.

D D Ebbing, *General Chemistry* (5th edition)
Houghton Mifflin, 1996

(a) *Which of the following statements are underlying assumptions in the passage?*

A Molecules of the liquid form stronger bonds with each other than with molecules of whatever substance is at the interface.

B The units of surface tension are $J\,m^{-2}$.

C Dissolved substances change the nature of interactions between molecules in the solvent.

D Water molecules interact with each other more strongly than they interact with the surface of a pin.

E Surface tension of a liquid is defined as being measured at an inter-face with air.

F Waterbugs float on pure water.

(b) *Write a short passage in which you explain in simple terms why the pin would float on top of the water and why it would sink if soap was added.*

6 Neil Bartlett, working at the University of British Columbia, prepared the first noble-gas compound after he discovered that molecular oxygen reacts with platinum hexafluoride, PtF_6, to form the ionic acid $[O_2^+][PtF_6^-]$. Because the ionisation energy of xenon (1.17×10^3 kJ/mol) is close to that of molecular oxygen (1.21×10^3 kJ/mol), Bartlett reasoned that xenon should also react with platinum hexafluoride. In 1962 he reported the synthesis of an orange-yellow compound with the approximate formula $XePtF_6$.

D D Ebbing, *General Chemistry* (5th edition)
Houghton Mifflin, 1996

(The product actually has variable composition and can be represented by the formula $Xe(PtF_2)_n$ where n is between 1 and 2.)

Which one of the following statements do you suppose best represents a step in Bartlett's reasoning that xenon should react with platinum hexafluoride?

A It is rather more energetically favourable to remove an electron from xenon than from an O_2 molecule.

B It is rather less energetically favourable to remove an electron from xenon than from an O_2 molecule.

C PtF_6 is a powerful oxidising agent.

D High activation energy normally prevents xenon from losing an electron.

7 If you want to know whether smoking damages your health, you cannot just say, 'My grandmother smoked 80 a day and is still alive'. It has to be done statistically. The world would be a better place if there were a few more people in Parliament, in the law and the civil service who were trained at least in statistics if not the rest of science.

R Dawkins, *The Bookseller*
April, 1996

(a) *Which one of the following statements, any of which might be made in relation to the above passage, do you agree with most strongly?*

A Statistics cannot tell you whether or not smoking will damage your own health.

B The passage has no relevance to you either because your grandmother is still alive or because she smokes less than 80 cigarettes a day, or both.

C There is no evidence that the world would be a better place if more people in Parliament, the law, and the civil services were trained in statistics.

D So few people are trained in statistics that a few more would not make any difference.

E More people ought to be trained in statistics.

F There is no proof that smoking damages health.

G Statistics can be manipulated so as to support virtually any argument.

(b) *Write a paragraph for a non-scientist reader which explains why you support the statement which you selected.*

8 For a particle moving in three-dimensional space, the square of the velocity is equal to the sum of the squares of its component velocities v_x, v_y and v_z and in the axial directions x, y and z. Its kinetic energy is given by

$$T = \tfrac{1}{2}mv_x^2 + \tfrac{1}{2}mv_y^2 + \tfrac{1}{2}mv_z^2$$

and the Schrödinger equation for a single particle in three dimensions is

$$\frac{\partial^2\psi}{\partial x^2} + \frac{\partial^2\psi}{\partial y^2} + \frac{\partial^2\psi}{\partial z^2} + \frac{8\pi^2 m}{h^2}(E - V)\psi = 0$$

Then $\psi^2\,dxdydz$ at any point (x,y,z) is a measure of the relative probability of finding the particle within the volume $dxdydz$ at (x,y,z). Multiplying the equation for ψ by a constant does not change the relative values of ψ at different points, and to maintain the concept of probability the equation obtained for ψ in a given system may have to be normalised, i.e. multiplied by a constant factor such that the integral of $\psi^2\,dxdydz$ over all space is unity.

A G Sharpe, *Inorganic Chemistry* (3rd edition)
Longman, 1992

Which one of the following statements best represents the important message in this passage?

A The kinetic energy of the particle is $\tfrac{1}{2}mv_x^2 + \tfrac{1}{2}mv_y^2 + \tfrac{1}{2}mv_z^2$

B Wave mechanics is a useful tool for predicting the probability of finding an electron in a particular volume of space.

C The relative value of ψ at different points is not changed by multi-plying equation for ψ by a constant.

D Schrödinger was a genius.

E The Schrödinger equation for a single particle in three dimensions is

$$\frac{\partial^2 \psi}{dx^2} + \frac{\partial^2 \psi}{dy^2} + \frac{\partial^2 \psi}{dz^2} + \frac{8\pi^2 m}{b^2}(E - V)\psi = 0$$

9 The EU wants methane and nitrous oxide to be included immediately in a 'basket' of greenhouse gases covered by the agreement [to set targets for max-imum emission levels].

The UN's Intergovernmental Panel on Climate Change (IPCC), which is pro-viding the scientific input to the climate talks, admits that its estimate of emis-sions of methane from human activity could be wrong by 20%, and of nitrous oxide by 60%. While estimates of methane emissions from coal mines, nat-ural gas pipelines and landfill sites are reasonably accurate, those from rice paddies and cattle are not.

Calculations of the long-term warming caused by a particular emission are complicated by the fact that the different gases have average residence times in the atmosphere ranging from 12 years to thousands of years. Measured over 100 years, the benefit to the world's climate of preventing the release of a given amount of methane is 21 times as great as not releasing the same amount of CO_2. But measured over 500 years, the benefit is only 6.5 times as great.

F Pearce, *New Scientist*
15 March, 1997

(a) *On the basis of the information presented here, which one of the following state-ments do you most strongly support?*

A The problem of global warming due to methane emission is greater than that of carbon dioxide, so that more attention should be given to methane emission.

B The estimation of methane emissions from rice paddies and cattle could be in error by ±20%.

C There is not enough information in the passage to be able to decide whether the emission of methane or nitrous oxide from human activity is a significant contributor to global warming.

D The EU are right to want to set limits on the emission of methane and N_2O resulting from human activity.

E The number of cattle should be regulated.

(b) *Write a paragraph for a non-scientist which explains why you support the statement which you selected.*

10 It is natural for an able and experienced analyst to believe that if, for example, he finds 10 ng/ml of a drug in a urine sample, then other analysts would obtain closely similar results for the same sample, any differences being due to random errors only. Unfortunately, this is far from true in practice. In many collaborative trials involving different laboratories, even when meticulously prepared and distributed aliquots of a single sample are examined by the same experimental procedures and the same types of instrument, the variation in the results often very greatly exceeds that which could rea-

sonably be expected from random errors. The inescapable conclusion is that in many laboratories a variety of substantial systematic errors, both positive and negative, is going undetected or uncorrected.

J C Miller & J N Miller, *Statistics for Analytical Chemistry* (2nd edition)
Ellis Horwood, 1988

Which one of the following statements best represents the main message in this passage?

A Experienced analysts are sometimes careless and so get irreproducible results.

B Sources of systematic errors in analysis are often very subtle and so can only be detected by comparison of results from different laboratories.

C Collaborative trials often reveal differences in the spread of results from different laboratories.

D One must expect different results from different laboratories employing the same procedures carried out by analysts of different experience and qualifications.

E Random errors are not shown up by collaborative trial, only systematic errors.

11 Chemists at the University of Rochester have 'tricked' the enzyme that synthesises DNA into accepting an unnatural base in the template strand of DNA. The mimic has the same shape as the base thymine, but unlike thymine, it is unable to form strong hydrogen bonds with thymine's complementary base, adenine. Nevertheless, the enzyme 'reads' the base as if it were thymine and inserts an adenine in the complementary strand with a specificity almost as high as that for thymine itself.

'Most biochemistry textbooks cite hydrogen bonds as the primary reason that DNA is copied accurately,' says chemistry professor Eric T Kool, who headed the research effort. But in this experiment, DNA is copied accurately even though the base completely lacks conventional hydrogen bonds.

R Rawls, *Chemical and Engineering News*
3 March, 1997

The structure of the unnatural 'base' (difluorotoluene) is shown below, together with that of thymine with its complementary base, adenine.

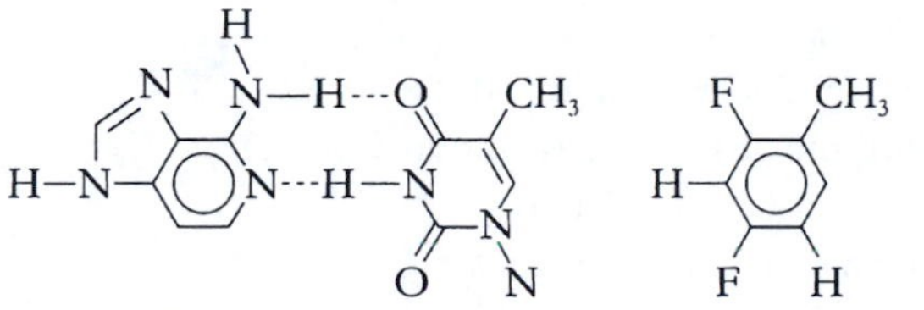

Adenine   Thymine   Difluorotoluene

In this context, which one of the following statements is the most appropriate comment on the similarity between difluorotoluene and thymine?

A   Fluorine is highly electronegative and so is like the oxygen in thymine in that it forms strong hydrogen bonds with suitable hydrogen donors.

B   Although only one of the oxygens in thymine forms hydrogen bonds with adenine, it was necessary to replace both with fluorine because thymine (or its analogue) can be rotated so that either of the oxygens (in thymine) can be used as the hydrogen bond acceptor.

C   Although fluorine is normally regarded as a good hydrogen bond acceptor, this is not the case when it is attached to an aromatic ring, since its lone pair of electrons can be potentially delocalised throughout the electron ring.

D   Difluorotoluene is aromatic, but thymine is not and so the former is not a good analogue of the latter.

12 | Another common failing in the logic behind odour theories is the inability to distinguish between cause and effect. All too often, someone finds a correlation between two parameters and assumes a causal relationship without

asking if this correlation could simply be between two effects of a common cause. An example could be the assumption that a correlation between the infrared spectra of a set of odorants and their odours demonstrates that the odour is caused by those specific vibrations. In fact, the odours and the spectra are both effects of a common cause, the molecular structure.

C Sell, *Chemistry in Britain*
March, 1997

This passage refers to a common failing in logic and gives an example. Which one of the following passages best represents:

(a) *The example of the failure in logic?*

(b) *A theory about odour which avoids the failure in logic referred to in the passage?*

A Only volatile molecules have an odour. Odorants have characteristic infrared spectra. Therefore the absorption of infrared radiation is a measure of volatility.

B 'Smell' is recognised because molecules bind to specific receptors. Molecules with similar molecular structure bind to the same receptors and therefore smell similarly. Molecules with similar structure also have similar infrared spectra. This explains why infrared spectra can be correlated with odour.

C Compounds with a similar smell show similar absorption bands in their infrared spectra. Infrared absorption is caused by molecular vibrations. Therefore smell is caused by molecular vibrations.

D Not all volatile molecules have an odour. There is a correlation between the infrared spectra of a set of odorants and their odours. Therefore volatile molecules which do not have an odour do not absorb infrared radiation.

13 Ascorbic acid (AA) is a powerful antioxidant and unlike many other antioxidants it has the ability to reduce molecular oxygen. It is widely used as an antioxidant in the food and beverage industry, and some brewers add AA at 30–50 mg l^{-1} to beer in order to diminish residual oxygen after packaging. This is thought to delay the development of stale flavours in beer. The sensitivity of AA to oxygen, however, makes the analysis of the reduced form very difficult, particularly if the sample must undergo an extraction technique that permits oxidation.

AA is very easily oxidized at a glassy carbon working electrode at low operating potentials, thereby providing a sensitive detection system. The ease of oxidation of AA means that potentially interfering analytes with higher oxidation potentials are not detected, so selectivity is high. While HPLC with electrochemical detection for AA determination has been widely documented, many reported methods suffer from either a lack of reproducibility or from losses of sensitivity due to fouling of the electrode with sample components. This has necessitated bracketing of samples and standards, which, as well as being troublesome to perform, greatly reduces the number of samples which may be analysed in a given time.

D Madigan, I McMurrough & M R Smyth, *Analytical Comms.*
1996, **33**, 9

This passage is taken from the introduction of a paper which describes a new method for determining ascorbic acid in beer.

(a) *Which one of the statements best reflects the message which the authors of the passage are trying to convey?*

A Ascorbic acid improves the keeping qualities of beer but, as a chemical additive, its use ought to be discouraged.

B There is a real need for a new method to determine the level of ascorbic acid in beer.

C Ascorbic acid is readily oxidised at a glassy carbon electrode and therefore electrochemical analysis provides sensitivity and selectivity.

D HPLC is widely used as a detection method for ascorbic acid.

E The use of HPLC for the analysis is slow because of the need to bracket samples with reference standards.

(b) *Explain in simple terms, understandable to the average beer drinker, why ascorbic acid is useful in the beer industry and why it is difficult to analyse.*

14 In recent years, much interest has been shown in the levels of transition metals in biological samples such as blood serum. Many determinations have been made of the levels of (for example) chromium in serum – with startling results. Different workers, all studying pooled serum samples from healthy subjects, have obtained chromium concentrations varying from <1 to ca. 200 ng/ml! In general the lower results have been obtained more recently, and it has gradually become apparent that the earlier, higher values were due at least in part to contamination of the samples by chromium from stainless-steel syringes, tube caps, and so on. The determination of traces of chromium, for example by atomic-absorption spectrometry, is in principle relatively straightforward, and no doubt each group of workers achieved results which seemed satisfactory in terms of precision, but in a number of cases the large systematic error introduced by the contamination was entirely overlooked. Methodological systematic errors of this kind are extremely common – incomplete washing of a precipitate in gravimetric analysis, and the indicator error in volumetric analysis are further well-known examples.

J C Miller & J N Miller, *Statistics for Analytical Chemistry* (2nd edition)
Ellis Horwood, 1988

Choose from the sentences below one which:

(a) *Best represents what this passage describes.*

(b) *Best represents what this passage exemplifies.*

A Trace levels of chromium are difficult to determine consistently by atomic absorption spectroscopy.

B Transition metals in biological samples may give false results by atomic absorption spectroscopy.

C In trace analysis, contamination from apparatus can be a serious problem.

D Chromium determinations are particularly liable to error due to leaching of ions from apparatus.

E Beware of systematic errors.

15 The passage below is a brief review of the mechanism by which ring closure takes place in the Pictet–Spengler reaction. The superscript numbers refer to other published papers which support the statements made.

There are two main pathways by which the ring closure could take place, involving either direct attack at the indole 2-position (route a), or attack at the 3-position followed by migration (route b).[2] Experiments on related systems have suggested that a spiroindolenine intermediate (4) is probably involved (route b);[3,4] however, electrophilic attack at the indole 2-position is known to compete (in acyclic systems) with attack at the 3-position,[5] and it has been noted that certain stereochemical features of the Pictet–Spengler reaction are consistent with either mechanism;[6] moreover, attack at the indole 3-position would presumably involve 'disfavoured' 5-*endo-trig* ring-closure, whereas direct attack at the 2-position could proceed through the 'favoured' 6-*endo-trig* pathway.[7]

P Bailey, *J. Chem. Research*
1987, 202

Which one of the following statements best represents the author's view of the most likely mechanism?

A Route a is the expected pathway.
B Route b is the expected pathway.
C Either route is plausible.
D Both routes take place simultaneously.
E Neither route is likely to be correct.

16 Red wine-making cannot use sulfur dioxide so effectively. Anthocyanin extraction requires contact between the juice and the grape skins, and this leads to contact of the juice with the high wild microorganism population on the skins. Even if sulfur dioxide is added, rapid growth of these microorganisms is difficult to avoid. The number and wide spectrum of microorganisms ensure that some will be comparatively tolerant to sulfur dioxide. These more tolerant microorganisms produce acetaldehyde (a precursor of ethanol) that binds sulfur dioxide as its bisulfite addition compound. As a result the sulfur dioxide concentration decreases inexorably until it becomes low enough to allow a wide range of microorganisms to proliferate. Attempts at control are futile, so sulfur dioxide is not used. Instead, a heavy dose of a vigorously growing, desirable yeast strain is added immediately after crushing to ensure the rapid onset of a fermentation in which most microorganisms are suitable.

M Allen, *Chemistry In Britain*
May, 1996

(a) *Which one of the following statements best describes why sulfur dioxide is not added during the making of red wine?*

- **A** Acetaldehyde produced during the fermentation of red wine destroys the sulfur dioxide.
- **B** Sulfur dioxide is not so effective at destroying the organisms in red wine as in white wine.
- **C** The juice from which red wine is made contains organisms which tolerate SO_2.
- **D** Contact of the juice with the large numbers and wide variety of organisms in grape skins during red wine fermentation renders SO_2 less effective than it is during white wine fermentation.
- **E** The sulfur dioxide is not able to control the undesirable organisms in red wine because of the heavy dose of yeast which is added to bring about rapid fermentation.

(b) *Explain in terms understandable to the general public why sulfur dioxide is not used in red wine making.*

17 Rabbit calicivirus appeared in Europe in the mid 1980s, and kills over 90% of infected rabbits in the early stages of an outbreak. Because rabbits are a major pest in Australia, a research programme was established to test the effectiveness of the virus as a way of controlling the rabbit plague. The programme was established on an island, but the virus escaped to the mainland in October 1995.

Smith, Matson, Cubitt and others say that to justify releasing the virus in the first place, the Australian government should have first obtained clear proof that it infects just one species, the rabbit. AAHL researchers claim to have done just that. Between 1991 and 1996, they exposed 31 species of native and domestic animal to the virus. Samples from two New Zealand species that had been exposed to the virus were sent to the AAHL for analysis. The researchers measured the amount of antibodies and virus in the blood and organs of these animals. They also looked for any signs of sickness. According to GSIRO's public statements, those tests showed that the virus did not replicate or cause disease in any test animal. 'Our testing of rabbit calicivirus is the most comprehensive study that we know of into the host range of an animal virus', says Murray.

A Anderson & R Nowak, *New Scientist*
22 February, 1997

(a) *Smith, Matson and Cubitt are said to have specified criteria which should be met before the virus should have been released; are the criteria reasonable?*

(b) *Do the tests outlined here provide clear proof that there is no danger from the rabbit virus to the species tested?*

(c) *Compare the statement attributed to Murray with the claim attributed to the AAHL researchers; are they consistent?*

(d) *Use this extract as the basis for a short article (1 or 2 paragraphs) for publication in a daily newspaper, taking into account your responses to (a), (b) and (c).*

18 This passage deals with some of the problems connected with the choice of carrier gas for use in a particular application of mass spectrometry. The application is known as inductively coupled plasma mass spectrometry (ICP-MS). This technique is particularly useful for the simultaneous determination of trace elements in a sample.

Although an argon ICP produces a high degree of ionization for most elements there remain a few which are inaccessible to it. These mostly have ionization energies over 10 eV, where the response is poor enough to be very susceptible to interference, or in a few cases to be negligible. In addition, massive peaks of the plasma and solvent gases completely obscure the major isotopes of some elements such as $^{40}Ca^+$ which coincides with $^{40}Ar^+$, $^{39}K^+$ which may be obscured by $^{38}Ar^1H^+$, and $^{80}Se^+$, coinciding with the argon dimer peak. Although analysis may be attempted using the minor isotopes of some of these elements, the resulting determination limits are proportionately poorer. The choice of other plasma gases is very limited and most attention has been paid to helium. This has a much higher ionization energy than argon (24.59, cf. 15.76 eV) so that a helium plasma should ionize all other elements well. Unfortunately, argon is one of these and is present as an impurity in most helium, and of course in any air ingested with the sample. Thus, argon interferences are usually present in such a plasma.

A Gray, *Analytical Comms.*
1994, **31**, 371

Which one of the statements below best summarises the passage?

A The determination limits of argon ICP are unsatisfactory for several elements.

B   Although the use of helium in place of argon should in principle allow the determination of presently inaccessible elements, in practice there are technical difficulties.

C   It is important to avoid getting air into the sample for ICP because the argon content of the air will cause interferences.

D   Ions of major isotopes of many elements including $^{39}K^+$, $^{40}Ca^+$ and $^{80}Se^+$ are obscured in argon ICP.

19 | In aqueous solution, protein molecules fold up into a compact globular structure in which most of the non-polar side chains are clumped in the core where they are shielded from contact with water. In this passage, Stryer explains why it is energetically unfavourable for a non-polar molecule to be surrounded by water – hence why it is energetically favourable for the non-polar side chains to clump together.

Consider the introduction of a single non-polar molecule, such as hexane, into some water. A cavity in the water is created, which temporarily disrupts some hydrogen bonds between water molecules. The displaced water molecules then reorient themselves to form a maximum number of new hydrogen bonds. This is accomplished at a price: the number of ways of forming hydrogen bonds in the cage of water around the hexane molecule is much fewer than in pure water. The water molecules around the hexane molecule are much more ordered than elsewhere in the solution.

L Stryer, *Biochemistry* (3rd edition)
W H Freeman & Co, 1988

Stryer gives no evidence to support this model; consider the following alternative model:

The non-polar molecule cannot form hydrogen bonds with water molecules. Therefore hydrogen bonds between water molecules are broken, and this has an unfavourable enthalpy term. The unfavourable enthalpy of dissolution of non-polar molecules means that it is favourable for the non-polar molecules to aggregate.

What experiment would you do to determine which of these is the better model?

20 | The principal use of hydrazine is as a rocket fuel. It is second only to liquid hydrogen in terms of the number of kilograms of thrust produced per kilogram of fuel burned. Hydrazine has several advantages over liquid H_2, however. It can be stored at room temperature, whereas liquid hydrogen must be stored at temperatures below −253 °C. Hydrazine is also more dense than liquid H_2 and therefore requires less storage space.

Pure hydrazine is seldom used as a rocket fuel, however, because it freezes at the temperatures encountered in the upper atmosphere. Hydrazine is mixed with *N,N*-dimethylhydrazine, $(CH_3)_2NNH_2$, to form a solution that remains a liquid at low temperatures. Mixtures of hydrazine and *N,N*-dimethylhydrazine were used to fuel the Titan II rockets that carried the Project Gemini spacecraft, and the reaction between hydrazine derivatives and N_2O_4 is still used to fuel the small rocket engines that enable the space shuttles to manoeuvre in space.

D D Ebbing, *General Chemistry* (5th edition)
Houghton Mifflin, 1996

(a) *What are the stated advantages of hydrazine as a rocket propellant?*

(b) *What are the disadvantages?*

(c) *Why is N_2O_4 specified as the oxidant, and not O_2?*

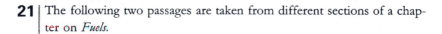

21 The following two passages are taken from different sections of a chapter on *Fuels*.

Another way of disposing of household rubbish containing polymers is to burn it. In several places household rubbish is made into pellets which can be used for solid fuel boilers. . . .

On heating addition polymers, the polymers melt very easily. It is called a thermoplastic polymer. When these start to burn they often produce highly poisonous gases, along with the carbon dioxide and water vapour that are produced in large quantities. Burning poly(acrylonitrile) for example can produce highly poisonous hydrogen cyanide.

R McDuell, *Revise Chemistry*
Letts, 1989

(a) *Why does the burning of poly(acrylonitrile) produce hydrogen cyanide when complete oxidation would produce H_2O, CO_2, NO_2?*

(b) *Would you consider it safe to burn poly(acrylonitrile):*

   (i) *On an open fire in a room in your house?*

   (ii) *In a solid fuel burner in a room in your house?*

   (iii) *In a furnace such as those used in power stations?*

(c) *Use these passages to write a paragraph which explains to a friend, who is not a scientist, the possibilities for and the problems of using waste polymers as a fuel. Take into account your responses to (a) and (b) and explain why poisonous gases may be produced and the circumstances which minimise this danger.*

22 | *Water molecules have a high affinity for each other.* A positively charged region in one water molecule tends to orient itself toward a negatively charged region in one of its neighbours. Ice has a highly regular crystalline structure in which all potential hydrogen bonds are made. Liquid water has a partly ordered structure in which hydrogen-bonded clusters of molecules are continually forming and breaking up. Each molecule is hydrogen bonded to an average of 3.4 neighbours in liquid water, compared with 4 in ice. *Water is highly cohesive.*

L Stryer, *Biochemistry* (3rd edition)
W H Freeman & Co, 1988

Which one of the following statements do you consider to be the best interpretation of 'each molecule is hydrogen bonded to an average of 3.4 neighbours'?

A Some molecules are hydrogen bonded to 3 others and some to 4 others; the average number is 3.4.

B Each molecule may be hydrogen bonded to any number of other molecules from 0–4; the average number is 3.4.

C Each molecule is hydrogen bonded to 4 other molecules, but not all of these are of ideal length and angle; the average bond strength of the 4 hydrogen bonds is equal to 3.4 ideal hydrogen bonds.

D The number of hydrogen bonds (which may not be ideal) formed by any single water molecule may be more than 4; the sum of the bond strengths of all the hydrogen bonds formed by each water molecule is equal to that of 3.4 ideal hydrogen bonds.

23 Formaldehyde has been described as 'an ubiquitous air pollutant that has been of great concern because of its adverse health effects'. Its concentration in the atmosphere can reach levels of around 50 ppb.

The passage below is the summary of a paper in which the authors suggest that tree planting can be an effective way of reducing this pollution since trees absorb formaldehyde vapour from the atmosphere.

To estimate the effect of tree planting on atmospheric formaldehyde, the absorption of formaldehyde by various tree species was examined. The absorption rates varied from 8.6 (Japanese black pine) to 137 ng dm^{-2} h^{-1} ppb^{-1} (Lombardy poplar) at 1000 µmol of photons m^{-2} s^{-1}, and the absorption rate increased in the following order: deciduous broad-leaved tree species > evergreen broad-leaved tree species > coniferous tree species. In experiments in which the light intensity was varied, a linear relationship between the formaldehyde absorption rate and the transpiration rate was observed for three tree species. From the results obtained from a simplified gas diffusive resistance model, we can conclude that formaldehyde is absorbed through the stomata, and is rapidly metabolized by three species. Even at a high concentration of about 2000 ppb, trees have the ability to absorb atmospheric formaldehyde for at least 8 h without any visible foliar injury. We conclude that trees in general could act as an important sink for atmospheric formaldehyde.

T Kondo, K Hasegawa, R Velida, M Ouishi, A Mizakami & K Omasa,
Bull. Chem. Soc. Japan
1996, **69**, 3673

(a) *Does the information provided in the summary convince you that:*

   (i) *Trees can remove formaldehyde from the atmosphere at a useful rate?*
   (ii) *Exposure of trees to formaldehyde for prolonged periods will not damage them?*

(b) *Identify any biological terms with which you are unfamiliar, and decide whether this unfamiliarity prevents you from drawing useful conclusions.*

24 In connection with the determination of the properties of mixed solvents, Meade *et al.* examined the interaction between dimethyl sulfoxide (DMSO) and mixtures of acetonitrile and methanol. Infrared spectroscopy (IR) is one of the techniques used. The following passage is taken from the experimental section of their paper in which a problem encountered in the measurement is discussed.

The S = O stretching vibration of DMSO occurred in the range 1000–1100 cm^{-1} and was partially masked by a methanol absorption band. Thus the IR measurements for DMSO were carried out using perdeuteriomethanol. While this reduced the absorbance due to methanol sufficiently for the DMSO band to be observed, it was not possible to quantify the absorbance with any precision in solutions rich in methanol.

M Meade, K Hickey, Y McCarthy, W E Waghorne, M R Symons &
P P Rastogi, *J. Chem. Soc., Faraday Trans.*
1997, **93**, 563

(a) *What made it difficult to measure the infrared spectrum of DMSO?*

(b) *What is perdeuteriomethanol?*

(c) *Why might it be expected to solve the problem?*

(d) *Why would you expect the interactions to be the same in deuteriomethanol-acetonitrile as in methanol-acetonitrile mixtures?*

(e) *The problem is stated to be only partly solved by the use of perdeuteriomethanol. In which cases is the measurement least satisfactory? Suggest a reason for it.*

25 Glucosinolate is an undesirable constituent of rapeseed. Plant breeders therefore look for varieties with low levels of this compound. The conventional method for measuring glucosinolate involves X-ray fluorescence (XRF). Measurement of absorbance in the near infrared (NIR) might provide a cheaper and simpler alternative.

To test this suggestion the glucosinolate content of a set of calibration samples was determined by the accepted XRF method. The NIR spectrum of the same samples was recorded, and the absorbance at an appropriate peak was measured. Dividing an XRF value by the NIR absorbance of the same sample, provides a conversion factor by which any NIR value can be multiplied to give a glucosinolate content. The mean conversion factor from all the pairs of calibration samples should give a good value for the glycosinolate content of any new sample on which an NIR reading has been obtained.

The graph shows how the usefulness of this conversion factor was tested. The glucosinolate content determined by the XRF method has been plotted against the values calculated from NIR absorbance.

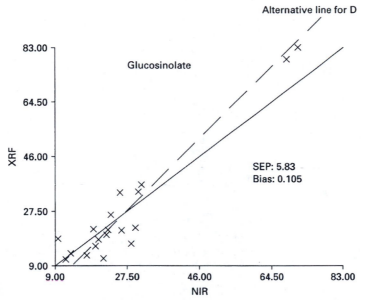

Plot of the glucosinolate content of samples of rapeseed as determined by X-ray fluorescence (XRF) and by near-infrared spectroscopy (NIR). (*From* A Salgó, Z Weinbrenner-Varga, Z Fábián & E Ungár, *Making Light Work*, 1992, I.M. Publications; reprinted by permission of I.M. Publications)

(a) *How do you suppose the authors justify the solid line shown?*

(b) *Which one of the following conclusions do you think can be most reasonably drawn from this graph?*

A   The results are so scattered that neither method is useful.

B   The graph proves that the calculated conversion factor is valid for all samples.

C   The data are not well enough distributed across the range to confirm that the conversion factor is valid.

D   The data would be better fitted by a line of steeper slope, as shown.

26 It is thought that because *para*-substituted aromatic molecules are more symmetrical than *ortho*-substituted ones, they will pack better and so have higher melting points. This old concept has been examined by comparing melting points of *ortho*- and *para*-disubstituted benzenes. The results are summarised in the graph.

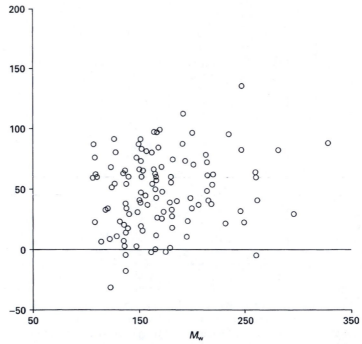

Differences in melting temperature between *para-* and *ortho*-disubstituted benzenes (*y* axis). The *x* axis is molecular weight. (*From* A Gavezotti, *J. Chem. Soc., Perkin II*, 1995, 328; reproduced by permission of The Royal Society of Chemistry)

Which one of the following interpretations of the graph is the most reasonable?

A The data are so scattered that no reliable conclusions can be drawn.

B There is no relationship between difference in the melting points and molecular weight.

C *Ortho*-substituted aromatics generally melt at lower temperatures than *para*-substituted aromatics.

D The difference in melting points increases with molecular mass.

27 | Zinc ions complex with the amino acid tyrosine and the complex is fluorescent. This has been developed into a method for the determination of zinc ions in solution. The authors of a paper on this procedure report that:

- The fluorescence intensity reaches a maximum when the molar ratio of tyrosine to Zn ions is 3; higher ratios have no effect on fluorescence.
- Fluorescence intensity reaches a maximum in a few seconds and remains constant for 1 hour.

The graph reproduced below describes the results of their study of the effect of sodium hydroxide on the system. The experiment was carried out by adding different volumes of NaOH solution (1.5×10^{-2} mol dm^{-3}) to a mixture of Zn^{2+} and tyrosine and making the solution up to constant volume. The final concentrations are given as Zn^{2+} = 4×10^{-4} mol dm^{-3} and tyrosine-0.06 mg cm^{-3}.

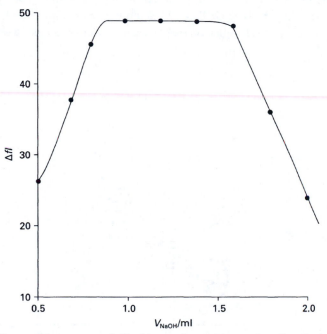

The effect on fluorescence (Δfl) of increasing volumes of sodium hydroxide in the reaction mixture. (*From* N Jie, J Yang, X Huang & W Ma, *Anal. Proc.*, 1995, **32**, 1399; reproduced by permission of The Royal Society of Chemistry)

Which one of the following is the most reasonable interpretation of the graph?

A The trace is really parabolic, but the line is hitting the top of the chart.

B The trace is really parabolic but the detector has become saturated so that a Δfl of 49 is the highest recordable value.

C The zinc-tyrosine complex begins to dissociate at high and low pH.

D The ratio of tyrosine to zinc is insufficient to produce the maximum fluorescence.

28 Triglycerides are esters of glycerol (1,2,3-propantriol) and long-chain aliphatic (fatty) acids. The composition of the fatty acids affects the suitability of the oil for some purposes. It is therefore useful to have a simple method for analysis. The standard method is gas chromatography. An alternative is near-infrared spectroscopy. The graph shown below is taken from a paper in which the two methods were compared. The method involved an initial calibration step in which a range of samples were analysed

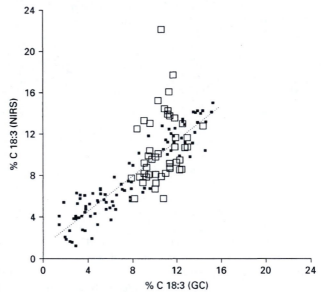

Percentage of linolenic acid (C 18:3) in different samples of rapeseed. The results from gas chromatography (GC) are plotted against those obtained by near-infrared spectroscopy (NIRS) for calibration samples (■) and validation samples (□). (*From* T C Reinhardt, C Paul & G Röbbelen, *Making Light Work*, 1992, I.M. Publications; reprinted by permission of I.M. Publications)

using both methods and the results used to calculate the line of best fit. In a second validation step the initial calibration step was checked using a further set of samples. Both sets of data are plotted on the graph.

Which one of the following statements can most usefully be made about the graph?

A If the two samples analysing at about 18 and 22% by near-infrared analysis were removed as outliers, the graph would look a lot better.

B Probably the best line through all the points is a curve rather than the straight line as drawn.

C The calibrations developed for the linolenic acid do not stand up to independent validation.

D The calibration and the validation samples cannot both be fitted by the same straight line.

E The positive intercept on the y axis shows that the near-infrared measurement gives spuriously high results.

29 One in ten of the world's population are facing starvation. There are more people alive today than have ever lived. Every hour we need an extra two and a half square miles of farmland to feed the world's increasing population.

Obviously, the problems of feeding all of the people of the world will be with us for a very long time. Scientists, especially chemists, have made and are making a significant contribution to solving these problems.

The solution of the problem is very complex. It is not just a matter of growing more food. Ways of improving the situation include:

5. Manufacture of new kinds of food. For example, cattle double their weight in 2 to 4 months while plants double their weight in 1 to 2 weeks. Yeast and bacteria double their weight in 20 minutes. A shallow lake the size of Essex could supply sufficient protein in the form of protein rich bacteria to supply the whole world's need for protein. Although we might not want to eat protein rich bacteria, it could be used for high protein animal feeds.

R McDuell, *Revise Chemistry*
Letts, 1989

The original passage includes nine possible ways of improving the world food problem. Only the fifth one is included here.

(a) *Suppose you put into a shallow garden pond a sample of water containing 1 μg of suitable bacteria.*

    (i) *How many doubling times must elapse before the mass of bacteria exceeds 100 mg?*

(ii) *Use the value you obtain to calculate an approximate value for the mass of bacteria you would expect to be present after 24 hours.*

(iii) *What kind of conditions do you suppose would be needed for bacteria to double their weight in 20 minutes?*

(b) *What information do you need in order to decide whether 'a shallow lake the size of Essex' could supply sufficient protein (as protein rich bacteria) to supply the world's need for protein?*

(i) *If the bacteria were processed directly into food for humans.*

(ii) *If the bacteria were used as animal feed.*

Data are given in the commentary which will enable you to do a calculation.

(c) *Write a paragraph for a non-scientist reader which explains the basis of the suggestion that the world food problem could be alleviated by growing bacteria in shallow lakes, and why this would not be seriously considered as a useful solution.*

30 The data shown in the graph below have been published in a paper about the catalytic activity of the enzyme chymotrypsin. A well-studied process is the catalysis by chymotrypsin of the hydrolysis of the ethyl ester of N-acetyltyrosine (ATEE). The rate of hydrolysis is inhibited by the presence of the ethyl ester of 3-iodo-N-acetyltyrosine (MATEE) or of 3,5-diiodo-N-acetyltyrosine (DATEE).

The rate (v) at which chymotrypsin hydrolyses ATEE is affected by the concentration ([S]) of ATEE in accordance with the equation:

$$v = \frac{V_{max}[S]}{K_M + [S]}$$

where V_{max} and K_M are constants.

The 'Lineweaver–Burke plot' is obtained from a linearised form of this equation:

$$\frac{1}{v} = \frac{1}{[S]} \cdot \frac{K_M}{V_{max}} + \frac{1}{V_{max}}$$

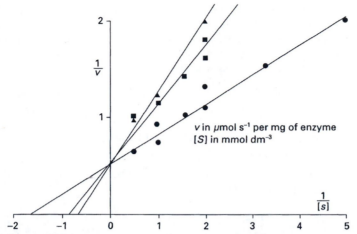

Lineweaver–Burke plots for the hydrolysis of ATEE (■) and in the presence 1.70 mM MATEE (●) and 0.55 mM DATEE (▲). (Data previously published in C J Garratt & D M Harrison, *FEBS Letters*, 1970, **11**, 17)

(a) *Does the figure convince you that the rate of hydrolysis of ATEE is reduced in the presence of either MATEE or DATEE?*

(b) *What principle(s) do you suppose the authors used in fitting the lines to the data?*

(c) *Do you agree that the lines are justified?*

31 The plant kingdom is an intermediate stage for genetic engineering; many people would accept that the development of new plant species in this way is little different, ethically, from trying to develop a new rhododendron by cross pollination, or even by irradiation which produces random genetic variation. But engineering the gene directly is different because, given full understanding of the function of the genes and the protein complexes which they form, almost anything becomes possible, in a controlled way. We might produce grass made largely of rubber or trees which spout petroleum like an oil well.

R Porter, *New Scientist*
20 November, 1986

(a) *In your opinion, what daily rate of oil production would a tree need to achieve in order for it to 'spout petroleum like an oil well'?*

(b) *Identify the information you need in order to estimate the maximum average hourly rate at which a tree could 'spout' oil (averaged over a year).*
     The commentary gives information which will enable you to check that you have identified what you need to know and to estimate an hourly rate of spouting.

32 The potato is the best package of nutrition in the world, rich in minerals, vitamins, calories and protein, and virtually fat-free. And potato fields yield more tonnes – and more calories – per hectare than any other form of cultivation, averaging 13.5 tonnes per hectare worldwide, four times as much as rice and five times as much as wheat.

F Pearce, *New Scientist*
26 April, 1997

The following data for the energy and protein content per 100 g of potatoes, flour (ground wheat), and rice are taken from *Manual of Nutrition*, London, HMSO, 1977:

	Energy/KJ	Protein/g
Potatoes	324	2.1
Wheat	1483	10.0
Rice	1531	6.2

(a) *Using the figures given in the passage, which crop gives:*

    (i) *The highest yield of energy per hectare?*
    (ii) *The highest yield of protein per hectare?*

(b) *On the basis of the information given here, do you think the passage is a fair comment on the food value of the potato?*

(c) *Use this passage and the data provided to write a paragraph for the average shopper which compares the food value of the potato with that of wheat and rice in terms of the land needed to grow them.*

33 The graph below is taken from a D.Phil thesis. It shows the rate at which deuterium (from D_2O) is exchanged for hydrogen in the 48 amide groups found in the protein insulin. The data are interpreted to mean that the amide hydrogens can be assigned to one of three groups.

- 27 exchange so fast that the exchange is complete before any observations can be made.
- 12 exchange with an effective first order rate constant of about 2.5 s⁻¹.
- 9 exchange with an effective first order rate constant of about 0.5 s⁻¹.

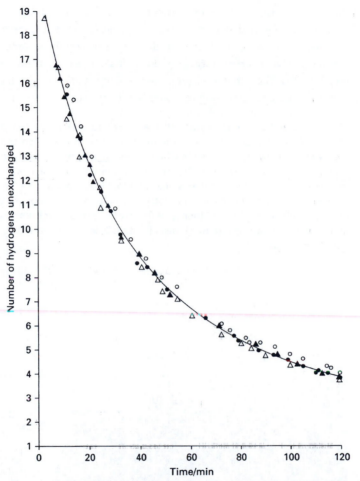

The time course of exchange of deuterium for amide hydrogens in beef insulin (different symbols represent different samples). (Data previously published in R A Capaldi & C J Garratt, *European Journal Biochemistry*, 1971, **23**, 551)

(a) *How would you check whether the data can be fitted by a first order plot?*

(b) *Assuming that the numbers of unexchanged amide hydrogens (y axis) are correctly determined, how would you use the figure to assess whether the authors conclusions are reasonable?*

34 Published statistical data rarely provide information in exactly the form you want. Published data therefore often need to be interpreted and processed. Here are data from two different sources about atmospheric carbon dioxide. Both the passage and the diagram are taken from articles which discuss the risk that major climate change may result from continued human activity releasing carbon dioxide into the atmosphere.

There is no doubt that man's activities are leading to a gradual increase in the atmospheric carbon dioxide level and this leads to the prospect that we may eventually modify the global climate. Fossil-fuel burning and deforestation are the main contributors to the global annual emissions, which have increased by a factor of about 10 since 1900 to an enormous 5.3×10^9 tonnes in 1980. This must be considered in relation to the total atmospheric content of CO_2 which is ca. 720×10^9 tonnes. If there were no removal processes, we should expect to be seeing an increase in the CO_2 level by about 0.75% of its value each year.

A G Clarke, *Understanding our Environment*, R E Hester (ed)
Royal Society of Chemistry, 1986

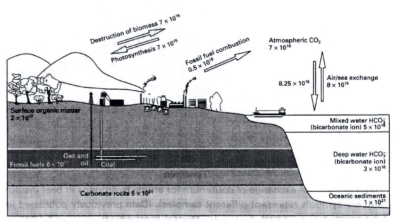

Representation of the major reservoirs and annual transfer rates involved in the global carbon cycle. The numbers are estimated numbers of moles of carbon held in specified forms. The numbers alongside arrows are estimates of the numbers of moles of carbon transferred from one reservoir to another in the direction indicated during the year. (*From* I Campbell, *Chemistry Review*, 1992, 1(4); © Philip Allen Publishers)

Compare the data given in the two passages for the production of CO_2 by fossil fuel burning and deforestation. In what way is one set of data misleading?

35 The passage below describes the measurement of the distribution of radio-active ^{14}C between the fatty acid and the glycerol component of triglyceride. A triglyceride is a triacyl ester of glycerol (1,2,3-propantriol). The acids with which the glycerol is esterfied are straight-chain alkyl carboxylic acids where the alkyl chain is typically 13–23 carbon atoms long.

Bray's scintillation fluid contains a compound which is excited by a radioactive disintegration and emits a photon as it returns to the ground state; the photon can be detected by a scintillation counter.

Triglycerides were extracted by washing the contents of the sample tubes into stoppered test tubes with 5 ml chloroform/methanol (2:1, v/v), mixing thoroughly with a vortex mixer and leaving to extract overnight. The chloroform layer was washed with chloroform/methanol/NH_2SO_4 (3:48:47, by vol.).

A sample of the washed chloroform layer was transferred to a counting vial, evaporated to dryness and redissolved in Brays scintillation fluid.

Another sample was dried and saponified by heating at 70 °C for 2 h with 0.5 ml M ethanolic KOH. The released fatty acids were extracted with heptane after acidification of the saponified sample. A sample of the heptane extract was counted.

R M Evans & C J Garratt, *Biochim. Biophys. Acta* 1977, **489**, 48

(a) *How did the authors determine the rate of radioactivity decay in the triglyceride?*

(b) *What is the process of saponification?*

(c) *Why did the chloroform/methanol used to extract triglyceride separate out into two layers?*

(d) *What was the volume of the washed chloroform layer?*

(e) *What is the meaning of 'a sample of the heptane extract was counted'?*

(f) *How did the authors determine the rate of radioactive decay in the glycerol component of triglyceride?*

(g) *Use the last two paragraphs to draw up detailed instructions for the processes described which would be suitable for incorporation into a laboratory manual used by undergraduate students.*

36 The following passage describes the preparation of reagents for two alternative procedures for measuring vitamin A. One is for a batch method of analysis, and the other involves flow injection analysis (FIA). The two antioxidants BHT and BHA are defined elsewhere in the paper as butylated hydroxytoluene and butylated hydroxyanisole.

Reagents preparation

The saponification solution of potassium hydroxide for the batch method was prepared by dissolving 50 g of KOH (pellets) in 100 ml of water; for the FIA method this was altered to 30 g of KOH in 100 ml of water.

The solution of ethanol used in the FIA method was prepared by taking 90 ml of absolute ethanol and diluting it to 100 ml with water. This solution was de-gassed for 15 min before being used in the FIA system, so avoiding bubble formation.

The antioxidant solution was prepared by dissolution of 25 mg of each of BHT and BHA in 100 ml of absolute ethanol. This solution was kept in a brown glass bottle in the dark, where it remained stable for several days.

E L Pérez & S J Haswell, *Anal. Proc.*
1995, **32**, 85

Describe carefully and precisely how you would carry out the three procedures given. Include details of the equipment, apparatus, and reagents which you would use.

37 | The following passage describes the preparation of an alicyclic *N*-sulfinyl compound. *N*-Sulfinylcyclohexylamine [5]:

Thionyl chloride (11.9 g) is added dropwise to a mixture of cyclohexylamine (9.9 g) and pyridine (7.9 g) in anhydrous benzene (50 ml), and the amine hydrochloride is filtered off. With the customary procedure, 9.5 g of thionyl-cyclohexylamine (65.4%) are obtained; bp 78 °C/15 mm.

G Kresze, A Maschke, R Albrecht, K Bedereke, H P Patzshke,
H Smalla & A Trede, *Angew. Chem. Internat. Ed. Eng.*
1962, **1**, 89

(a) *Expand this into a detailed recipe suitable for inclusion in an undergraduate laboratory manual. Include details of the apparatus and equipment required.*

(b) *Suggest change(s) which you would make on the grounds of safety.*

MAKING
JUDGEMENTS

Chemistry is often seen as a subject in which all questions have a 'right answer', and that the student's job is to learn what this right answer is. The exercises in this section are designed to show that this is an over-simplified view, and that many problems faced by chemists involve making a judgement about what kind of answer is acceptable. Some of the exercises are concerned simply with the meaning of words. As chemists we sometimes give special technical meanings to words which may mean something different in everyday conversation (for example 'organic'). We need to recognise when we are using the words in the technical sense. But we also need to recognise that, as chemists, we often use technical words (like 'solubility') which may not have a precise meaning.

The practicing chemist is faced with a wide range of questions to which there is no single right answer. For example, what is the best method by which to make a particular measurement? How do you best plan an experiment to test a particular hypothesis? Is there only one possible way to interpret experimental data?

The exercises in this section are designed to bring to your attention some examples of questions which do not have a right answer. Some of them involve doing a rough 'back of the envelope' calculation. In these exercises you will almost certainly not have enough information to carry out a precise calculation; the question you have to ask yourself is 'does it matter?' You may be surprised to discover how often you can get useful information from this kind of calculation; when you are planning an experiment or evaluating evidence, it is often useful to estimate an answer which is likely to be within one or two orders of magnitude of the right one.

Each sub-section in this section deals with an identifiable aspect of chemistry, but the style of exercise within each sub-section is very variable.

A key point of these exercises is to encourage discussion of questions to which we believe there is no single correct answer. It follows that our commentaries provide some guidance on an approach to thinking about particular exercises, but we do not give 'preferred answers'. The commentary on Exercises 1 and 2 in the first section provides an example of this.

SOLVENTS AND SOLUBILITY

Many chemical reactions occur in solution, and many procedures involve the use of solvents. Therefore, chemists need to use their knowledge of the factors affecting solubility to make judgements about the suitability of solvents and solutions for different purposes. This involves using knowledge and understanding of factors affecting solubility. It also involves recognising that the language we use is often not particularly precise. Chemists need to be able to interpret the meaning of some key words according to the context in which they are used

1 (a) Use the following questions to discuss whether you would describe calcium carbonate as soluble.

   (i) Are sea shells soluble?
   (ii) Would you consider building your house out of calcium carbonate?
   (iii) Where does the scale in kettles come from?
   (iv) How are stalactites formed?

(b) Use the following questions to discuss why there is concern that lead from old water pipes dissolves in drinking water.

   (i) Is it possible to measure the concentration of lead in water?
   (ii) Is lead soluble?
   (iii) What lead salts are soluble in water?

(c) Trichloromethane (chloroform) and diethylether both cause anasthesia if inhaled.

   (i) Do you regard either or both of those substances as soluble in water?
   (ii) How do these substances get from the lungs to the brain?

(d) The solubility product for $Fe(OH)_3$ is 10^{-39}. Concentrations of Fe^{3+} in river water are often quoted as being approximately 10^{-6} mol dm^{-3}. How can these two statements be reconciled?

(e) What do you mean when you describe something as

   (i) soluble?
   (ii) sparingly soluble?
   (iii) insoluble?

2 In the light of your answers to the following questions, discuss what you mean by a concentrated solution.

(a) Is a saturated solution always a concentrated solution?

(b) Is a 1% solution of sodium chloride (table salt) more or less concentrated than a 1% solution of sucrose (sugar)?

(c) Concentrated sulfuric acid must be handled with care; to dilute it, it must be added slowly to water; how dilute must it be before you regard it as no longer concentrated?

(d) Does your answer to (c) give you a rule of thumb which you can apply generally to define what you mean by concentrated?

GETTING STARTED

Familiar words like 'soluble', 'concentrated', and 'dilute' are useful even though they can have no precise chemical meaning. To answer these questions you need to think about the way we use words like this, the value of having words the meaning of which depends on the context, and some of the ambiguities which may arise if we are not careful. Exercise 1(b) reminds you that chemical names may themselves be ambiguous; we often use the word 'lead' to mean either the metallic or ionic form. Exercise 1(d) reminds you that substances described as being in solution may have been complexed by another solute; you may like to think about the difference in principle (if there is one) between an Fe^{3+} ion complexed to (say) a dissolved tannin and a sodium ion solvated by water. Exercise 2(b) reminds you that circumstances may determine the basic frame of reference; in this example, you need to decide whether you are more concerned with measuring the amount of solute as a mass or a number of moles.

3 | Given the following equipment:

* an ice bath
* a boiling water bath
* $HCl(aq)$
* $KCl(s)$
* $NH_4OH(aq)$

how would you

(a) maximise

(b) minimise

the solubility in water of each of the following compounds.

(i) Sodium carbonate

(ii) Magnesium chloride

(iii) Pentanol

(iv) Palmitic acid

(v) Phenol

4 **(a)** What criteria would you apply when choosing a suitable solvent for recrystallising a solid product?

(b) If, on an initial analysis, two solvents appeared to be equally suitable for recrystallising a particular solid, how would you decide which to use?

5 **(a)** What criteria would you use to select a suitable solvent for the liquid-liquid extraction of a solute from an aqueous solution?

(b) Would you ever expect a solvent extraction to give complete extraction?

(c) How do you decide how many repeat extractions are necessary?

6 Suppose you wish to measure the solubility of a compound. What would determine the length of time over which you would leave the solute and solvent in contact before deciding that the solution was saturated?

SAFETY

Safety and safe practice is an essential aspect of laboratory work. Chemists routinely have to make judgements about the precautions which are needed when carrying out a particular procedure or using a particular substance. Sometimes it is helpful to put these into perspective by thinking about similar activities which we carry out as ordinary people outside the laboratory. The first two questions in this section are more or less quantifiable; the second two are very much matters of judgement. We recommend that you use your answers to these questions as the

basis for a discussion of what you mean by a 'safe procedure', and whether you regard any procedure as 'perfectly safe'.

1 What are the hazards involved in carrying out the following procedures?

(a) An ether extraction

(b) Evaporating ether to dryness

(c) Refluxing a sample on an oil bath

(d) Deep frying potatoes

(e) Heating a sealed sample to 500 °C

(f) Using a pressure cooker

(g) Recrystallising a sample from acetone

(h) Cleaning brushes with paint remover

(i) Carrying out a synthesis to incorporate an atom of ^{14}C into a compound

(j) Carrying out a synthesis to incorporate an atom of ^{131}I into a compound

2 What hazards are involved in the use of the following?

(a) Hydrogen peroxide solution

(b) Ether

(c) Phenols

(d) Inorganic perchlorates

(e) Ammonium nitrate

(f) Benzene

(g) Sodium hydroxide pellets

(h) Potassium cyanide

(i) Ethanol

(j) Pyridine

3 Should any of the following products be restricted or banned on safety grounds?

(a) Oven cleaner

(b) Dishwasher powder

(c) Creosote

(d) Glue

(e) Fireworks

(f) Barbecue lighter fluid

(g) Nitrate fertilisers

(h) Hair dyes

(i) Vodka

(j) Methylated spirits

4 If you were following advised procedures and safety precautions whilst working with tetrachloromethane (carbon tetrachloride) in a fume cupboard, would you be likely to be inhaling more unpleasant or dangerous vapours than when experiencing the following?

(a) Walking in a city street during rush hour.

(b) Filling your car with petrol.

(c) Attending a firework display.

(d) Sitting in the smoking compartment of a railway carriage.

REACTIONS

A great deal of chemistry is concerned with the study and the representation of chemical reactions. As with the previous two sections this involves the use of language the meaning of which may have to be interpreted according to the context. An additional feature which can give rise to ambiguity is the representations of structures and of reactions; you may need to consider how closely these represent the essential features of reality. When studying reactions, the need for making judgements extends beyond that of language and representation; real decisions have to be taken about methods. Even in situations where experienced chemists would agree on the best methodology, it is likely that each individual would make slightly different decisions about its application. The exercises in this section encourage you to explore these decisions at a number of different levels.

1 **(a)** In what context would you regard each of the substances listed below as unstable?

AND I SAY IT IS STABLE....

  (i) CF_2Cl_2
  (ii) NH_4NO_3
  (iii) Fe
  (iv) I_2
  (v) CH_4
  (vi) C_2H_6
  (vii) $SiCl_4$

(b) In the light of your answers, what do you mean when you describe something as

  (i) stable?
  (ii) unstable?

2 **(a)** Consider the following radioactive isotopes:

  • ^{40}K (half life 1.28×10^9 years)
  • ^{14}C (half life 5730 years)
  • ^{125}I (half life 60 days)
  • ^{131}I (half life 8.0 days)

  (i) In what context(s) would you regard them as decaying at a significant rate?
  (ii) In what context(s) would you regard the rate of radioactive emission as constant with time?

(b) What principle(s) would you use in deciding how many days you would store a sample of ^{131}I before it would be reasonable to assume that essentially all of the radioactivity had decayed away?

3 **(a)** In what sense do the equations given below not represent adequately what happens when the reactants are mixed under appropriate conditions for them to react?

(i) $CH_4(g) + H_2O(g) \rightleftharpoons 3H_2(g) + CO(g)$

(ii)

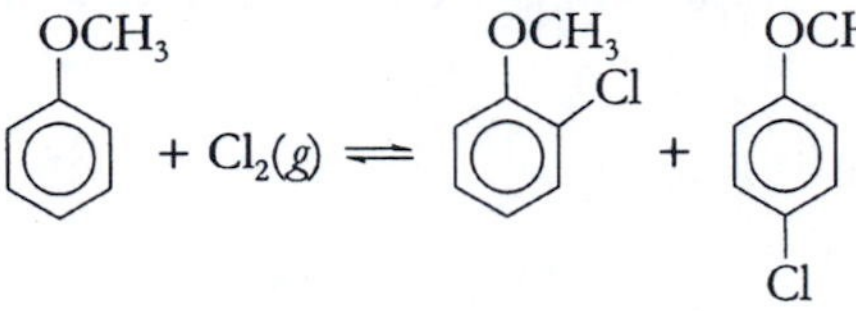

(iii)

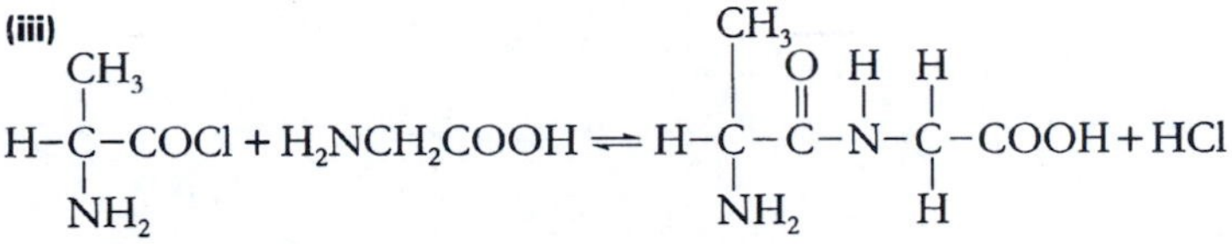

(iv) $CH_3CH=CHCH_3 + HCl \rightleftharpoons CH_3CH_2CHClCH_3$

(v)

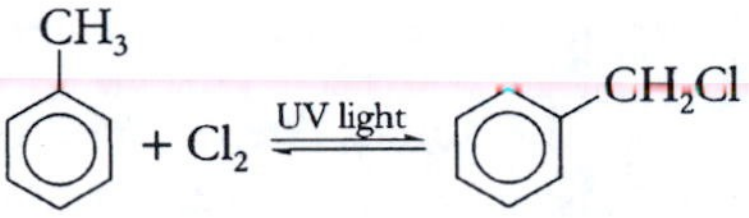

(b) Do chemical equations ever represent reality?

4 A scientist intends to compare the effectiveness of various substances as catalysts for the hydrolysis of an ethyl ester. Literature information indicates that a reasonable estimate of the rate could be made in 10–30 mins. Give reasons for preferring **one** of the following methods to follow the rate:

(a) Take samples of the reaction mixture at intervals and use GC to determine the concentration of ethanol.

(b) Follow the change in absorbance at 220 nm given that this is known to change as hydrolysis occurs.

(c) Measure the change in pH as the reaction proceeds.

(d) Measure the rate of addition of $NaOH(aq)$ needed to maintain a constant pH.

5 Crystal violet is a dye with a strong absorption in the visible region due to the extensive delocalisation in its structure. At a wavelength of 590 nm a solution of 10^{-5} mol dm^{-3} would give an absorbance of about 0.5 in a cell of 1 cm pathlength. It reacts with OH$^-$ to give a colourless product. The reaction can be represented as:

$$CV^+ + OH^- \rightarrow CVOH$$

where CV$^+$ represents structure A, and CVOH represents structure B.

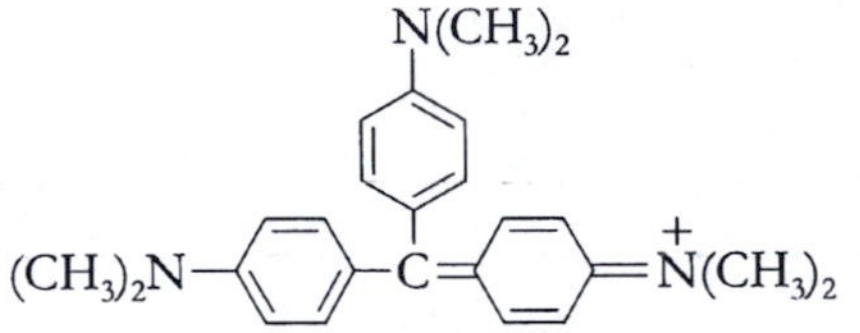

Structure A

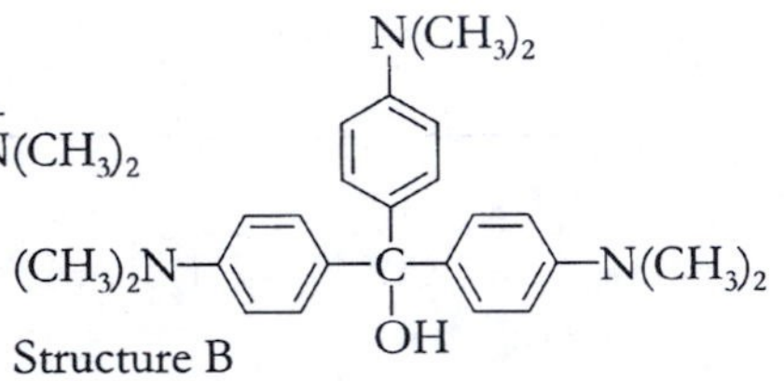

Structure B

From the equation you would expect the reaction to be first order with respect to each reactant, or second order overall.

(a) In order to obtain a pseudo first-order rate constant (k_{obs}) for the reaction, one of the reactants must be in excess. You can measure the change in concentration of crystal violet from the reduction in absorbance, and the rate of reduction of OH$^-$ from the change in pH.

  (i) Which reactant would you keep in excess, and why?

  (ii) What minimum ratio would you require for the reactant 'in excess' to that of the other reactant?

  (iii) Assuming you were unable to make a continuous recording of either absorbance or pH, how many measurements of reactant concentration would you make in order to obtain a pseudo first-order rate constant (k_{obs})?

(b) How would you determine the second order rate constant?

6 | The absorption spectra shown are those of a particular phenol (tyrosine) dissolved in dilute HCl and dilute NaOH

(a) Suppose you were required to study the pK of ionisation of tyrosine by measuring the absorption at different pH values. At what wavelength would you measure the absorption?

(b) Suppose you wished to measure the rate at which the phenol re-
acted with ICl to give an iodinated derivative. At what wavelength
would you make measurements in order to follow the reaction rate?

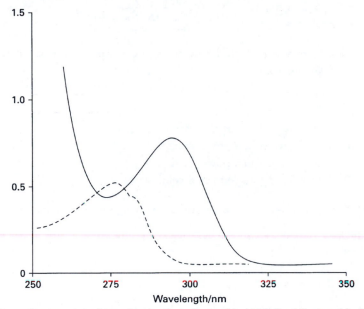

Absorption spectra of tyrosine (a phenol) at acid pH (dashed line) and in 0.1
mol dm KOH (solid line)

PURITY

Practicing chemists frequently have to make judgements about whether a
substance is sufficiently pure for a particular purpose. Since modern
methods of analysis are so sensitive, it is possible to show the presence
of impurities where once they would have gone undetected. One might
reasonably ask whether there can be such a thing as a pure substance. The
realist is more likely to ask whether it is pure enough. Most of the exercises in this sub-section are concerned with this point, and with the changes
in the meaning of words like 'pure', 'impure', 'contaminated' in different
contexts.

1 Choose from the list below

- Sea water
- River water
- Tap water
- Deionised water
- Distilled water

the type of water which, if any, would be acceptable for use in each of the following cases.

(a) Drinking

(b) Keeping fish

(c) Making up standards for trace metal analysis

(d) Cooling water in a power station

(e) Turbine water in a power station

(f) Water for injectable pharmaceuticals

2 What pH would you expect of the following samples of water?

(a) Sea water

(b) Tap water

(c) Distilled water

(d) Pure water

3 Which of the following impurities would you consider to be contaminants if present in a sample of drinking water?

(a) 150 ppm calcium ions

(b) 150 ppm lead ions

(c) 0.05% nitrate

(d) 5% ethanol

(e) 100 ppm fluoride

(f) 1 ppb benzene

4 In the light of your answers to the following questions, discuss what you mean by 'pure'.

(a) Is the composition of 'pure orange juice' always the same?

(b) Do you regard either table salt or sea salt as being

(i) pure?

(ii) pure sodium chloride?

(c) Sodium hexacyanoferrate(II) may be added to salt at a level of 0.01% to improve pouring qualities. Is this a contaminant?

(d) Wheat flour is made by grinding wheat grains.

  (i) Is wheat flour pure?
  (ii) Is white flour more or less pure than wholemeal flour?
  (iii) Chalk may be added to white flour to provide a source of calcium which is needed for bone development. Is this a contaminant?

(e) Which of the following types of ethanol would you regard as the purest?

  (i) That made by fermentation followed by fractional distillation to obtain 95% pure product.
  (ii) That made from 95% ethanol by distillation over benzene to give 99.8% ethanol.
  (iii) Denatured ethanol.
  (iv) Ethanol made from ethene.

(f) Consider whether you would regard the following substances as impure as a consequence of the accidental addition of 1% water?

  (i) Ethanol
  (ii) $SiCl_4$
  (iii) CCl_4
  (iv) NaOH
  (v) P_2O_5

YIELD

This is another sub-section where the most obvious need for judgement is in the meaning attached to a phrase like 'a good yield'. In some cases it may be sufficient to recognise what 'good' is likely to mean in a particular situation. But in some (real) situations it may be necessary to assess in advance what minimum yield will be acceptable, or to choose between different processes where the highest yield (of the key reaction) does not necessarily give the

most cost-effective yield (of the final product). These different aspects are covered in different exercises within this section.

1 On what grounds would you argue that a yield of 100% cannot be achieved?

2 The following passage describes a method for the synthesis of tin(IV) tetraiodide.

Tear 0.8 g of tin foil into small fragments and place in a dry round bottom flask with 3.8 g of iodine and 15 cm³ of 1,1,1-trichloroethane. Fit with a water-cooled condenser and reflux gently until all the tin has been consumed. With the condenser still in place, cool the flask in a water-ice bath. When the flask is cool, remove from the ice and water and quickly filter off the product through a dry sintered-glass crucible. Wash the product with cold, dry 1,1,1-trichloroethane and transfer to a vacuum dessicator and evacuate for 5 minutes. Recrystallise the crude product from 1,1,1-trichloroethane.

You are told that the maximum yield you can expect is 60%. When carrying out the experiment, you obtained a yield of

(a) 70%

(b) 40%

In each case suggest reasons for not obtaining the expected yield, and identify steps of the preparation to which you would pay particular attention when repeating the experiment.

3 In the synthesis of butyl propanoate, how would you increase the yield of the ester with respect to *n*-butanol?

$$n\text{-}C_4H_9OH + C_2H_5COOH \rightleftharpoons C_2H_5COOC_4H_9 + H_2O$$

4 Under what circumstances might you regard a yield of

(a) 40% as acceptable?

(b) 80% as acceptable?

(c) 95% as unacceptably low?

5 The scheme below is a diagrammatic representation of the Merrifield method for solid-phase peptide synthesis (R B Merrifield, L D Vizioli & H G Boman, *Biochemistry*, 1982, **21**, 5020).

Protected amino acid[1]   ■—N—C—C + Cl—CH₂—⬡—● Reactive resin

with structure:

$$\text{■—N—C—C}$$ (H, O, H, R¹, O⁻)

| Anchor

■—N—C—C—O—CH₂—⬡—● (H, O, H, R¹)

A | Deprotect with $F_3C–COOH$

N—C—C—O—CH₂—⬡—● (H, H, O, H, R¹)

B | Couple protected amino acid[2] with DCC

■—N—C—C—N—C—C—O—CH₂—⬡—● (H, O, R², H, O, R¹)

C | Sequential deprotection and coupling steps

D | Release with HF

H—N—C—C ⸜⸜⸜ N—C—C (H, H, O, Rⁿ, H, R¹, O⁻)

In this process the first amino acid is attached by its carbonyl group to a solid support; the amino group is protected by *t*-butyloxycarbonyl (tBOC).

Successive amino acids are added by a sequence of reactions as follows:

A The protecting tBOC is removed by treatment with trifluoroacetic acid.

B The tBOC-protected amino acid to be attached is added together with dicyclohexylcarbodiimide (DCC). DCC reacts with the carboxyl group to give an acyl derivative which is attacked by the amino group of the previous amino acid. The by-product from the DCC is removed.

C The process is repeated for each succeeding amino acid.

D The final peptide is cleaved from the solid phase with hydrofluoric acid (HF).

(a) **(i)** Suppose the yield of the coupling step was 95%, and of the other steps it was 100%. In an attempt to synthesise a tetra-peptide (i.e. four amino acids attached to each other) how many different molecules would be attached to the solid-support after the fourth amino acid had been allowed to react?

(ii) How many would be present in significant amounts?

(b) What would you regard as an acceptable yield for each step in this process, assuming you wanted to make a peptide 20 amino acids long?

6 Suppose you want to make

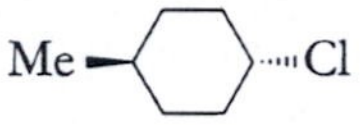

Me—⟨ ⟩····Cl

You have available as a starting material

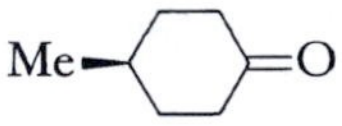

Me—⟨ ⟩=O

The proposed synthesis (shown below) involves:

• Reduction to an alcohol.
• Conversion of the alcohol to the chloride.

The initial reduction leads to the formation of the *cis* and *trans* diastereo-isomers. The overall yield of alcohol and the proportion of each isomer in the product can be altered by choice of reaction conditions.

The method for converting the alcohol into the chloride can be chosen either to retain or to invert the stereochemistry of the alcohol. This means that the desired product can be made from either of the pure diastereoisomers.

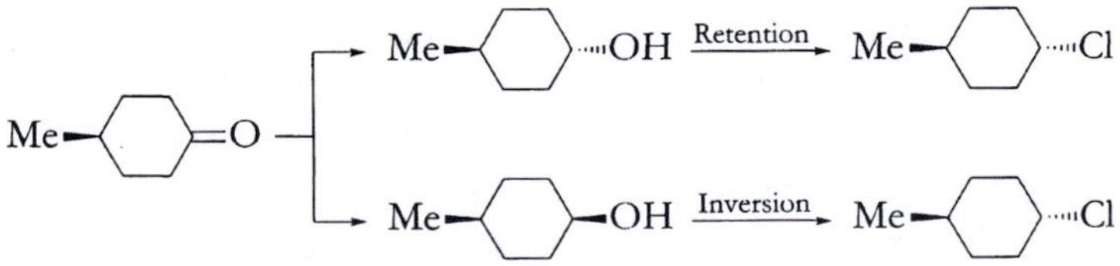

Me—⟨ ⟩=O ⎡ → Me—⟨ ⟩····OH $\xrightarrow{\text{Retention}}$ Me—⟨ ⟩····Cl
⎣ → Me—⟨ ⟩—OH $\xrightarrow{\text{Inversion}}$ Me—⟨ ⟩····Cl

Suppose you could select conditions for the reduction step which gave *either*

- An overall yield of alcohol of 50%, of which 95% is the *cis* isomer.

or

- An overall yield of alcohol of 95%, of which 50% is the *cis* isomer.

For each of the following four situations, which set of conditions would you choose?

(a) About 5 g of product is required.

  (i) A simple published procedure exists for separating the isomers.
  (ii) There is no known published method for separating the isomers.

(b) About 1000 kg of product is required.

  (i) A simple published procedure exists for separating the isomers.
  (ii) There is no known published method for separating the isomers.

7 In order to study the mechanism of the reaction between the indole derivative (I) and methanal (formaldehyde), the methanal was labelled with ^2H and the distribution of ^2H in the product (II) was determined. The reaction was carried out in methanolic HCl.

The required product was purified in 26% yield from the reaction mixture which also contained (III) in 50% yield.

Do you regard 26% as an acceptable yield for the purpose of this study of reaction mechanism?

8 A few decades ago the structures of all natural products were determined by a two-stage process. First the purified substance was degraded step-by-step and the structure deduced. The deduced structure was verified by synthesising the supposed substance from known materials and comparing the natural and synthesised product. A 20-step synthesis with an overall yield of 1% was not uncommon.

Is it reasonable to regard the synthesis as proof of the correctness of the structure when the procedure yields only 1% of the correct material?

ATOMS AND MOLECULES

In trying to understand chemistry and to predict the behaviour and property of chemicals we rely greatly on the atomic theory of matter. It is all too easy to forget that the variety of words, symbols, diagrams, and mathematical models which we use to represent atoms, molecules and subatomic particles are really only models which help us to represent some aspect of reality. It requires judgement to decide on the most appropriate way to represent these concepts or to describe the characteristics of particular substances. The exercises in this sub-section are designed to help you to recognise the need to make judgements when dealing with these fundamental concepts.

1 **(a)** Consider the following ways of representing a hydrogen atom and discuss how you would choose the best way to represent an atom.

- **(i)** H
- **(ii)** $_1^1$H
- **(iii)** $_1^2$H
- **(iv)** A white circle
- **(v)** A black circle
- **(vi)** A white circle with a stick attached
- **(vii)** A white sphere
- **(viii)** A white hemisphere
- **(ix)** The Schrödinger equation

(b) Since electrons are very small relative to the nucleus of an atom, in what sense does a 'space-filling' model of an atom represent space which is filled?

2 **(a)** **(i)** What do you mean by the radius of an atom of helium?
(ii) Do all atoms of helium have the same radius?
(iii) Does the radius of a particular atom of helium ever change?

(b) In answering the same questions about hydrogen instead of helium, what additional factors would you take into account?

3 In the light of your answers to the following questions, discuss what you mean by a molecule of water.

(a) **(i)** Does absolutely pure liquid water consist of molecules of H_2O?
(ii) In a molecule of H_2O do the same atoms of oxygen and hydrogen remain together for a significant period of time?
(iii) Does your concept of a 'molecule' of water depend on whether it is in the gaseous, liquid, or solid state?
(iv) Can you distinguish discrete molecules of H_2O in ice?

(b) What is the best way of representing the formula of aluminium trichloride?

(c) Could a molecule of 2-aminoethanoic acid (glycine, H_2NCH_2COOH) exist

(i) in aqueous solution?
(ii) under any other conditions?

(d) Is there such a thing as a molecule of

(i) NaCl
(ii) SiO_2
(iii) Nylon

4 In the light of your answers to the following questions, discuss whether you can define a metal unequivocally.

(a) Do you associate the elements in Group 14 or Group 15 with metals?
(b) What are the characteristics of metals?
(c) Can non-metals have metallic character?
(d) Can metals have non-metallic character?

5 | Valence bond theory is widely used to describe the nature of the bonding in simple molecules and ions. Are each of the examples given below adequately described by valence bond theory or is some other model required?

$$CO \qquad Co(CO)_6 \qquad C_2H_6 \qquad SO_4^{2-}$$

ACCURACY AND PRECISION

Chemists make many different kinds of measurement and need to be confident in the validity of these. The words 'accuracy' and 'precision' are important in this context. In ordinary usage they are virtually interchangeable. In science this is not so. Accuracy refers to the closeness of the result to the true value and is dependent on systematic error (or the lack of it). Precision refers to the spread of results and is determined by random error.

In science it is rarely (if ever) possible to have either complete accuracy or complete absence of random error. It is therefore always necessary to make a judgement about the level of accuracy and precision which is acceptable in any given situation. This means that it is important to have some 'feel' for the kind of accuracy and precision which can be obtained from different procedures.

The exercises in this sub-section are designed to help you think critically about these concepts in different contexts.

1 | How would each of the following affect your confidence in the accuracy of a measurement?

(a) You double the number of determinations.

(b) You use an additional method to calibrate your instrument.

(c) You get another chemist to repeat the measurement using the same procedure.

(d) You use two different procedures to obtain the value under consideration.

2 Suppose you require 50 cm³ of a standard solution containing approximately 2% of solute. You decide to use a top pan balance on the bench which displays figures to 0.0001 g.

(a) How accurately would you expect to know the mass of solid you have weighed?

(b) How accurately would you expect to know the concentration of the solution? (Express your answers in terms of the percentage error you expect.)

(c) In quoting the percent concentration of the solution, how many decimal places would you give?

(d) How would you change the procedure to increase your confidence in the true value of the concentration?

(e) Would your changed procedure result in greater precision, greater accuracy, or both?

3 Suppose you are required to use an accurately known amount of diethyl ether (about 10 cm³).

(a) Would you measure it by volume, or weigh it?

(b) Describe the procedure you would use.

(c) What confidence would you place on the value you obtain?

4 Volumetric glassware comes from the manufacturer already calibrated.

(a) To within what percentage of the correct value does each of the following have to be calibrated for routine analytical work?

  (i) A measuring cylinder
  (ii) A volumetric flask
  (iii) A volumetric pipette
  (iv) A graduated pipette
  (v) A burette

(b) What if anything do you need to know about the precision of the calibration method?

(c) Suggest a procedure for calibrating each of the pieces of glassware specified.

I THINK YOU'LL FIND IT'S 0.03ml SHORT

5 An experienced analyst and an experienced chromatographer are each asked to provide an analysis of an aromatic amine. The analyst obtains, by titration with acid, a value which is 99.8% of the theoretical value for the pure substance. The chromatographer reports that impurities amounting to 0.5% of the sample weight were detected.

Suggest possible reasons for the differences between these values and propose methods which would allow you to test your hypothesis.

6 Under the 1998 Road Traffic Act (in UK legislation) it is illegal to be in charge of a motor car if you have more than 80 mg of alcohol in 100 cm^3 of your blood. Suppose that the mean of **four** determinations made on blood taken from a driver gave a value of 81.0 mg per 100 cm^3.

Would you judge the driver to be over the limit?

7 The limit of detection of trace impurities is usually defined in terms of the signal-to-noise level (S/N) observable on a chart, for example, the ratio of the height of a chromatographic peak to the base line noise (see diagram).

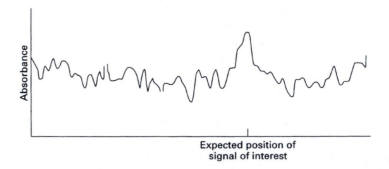

Expected position of
signal of interest

(a) How would you choose the base line?

(b) How would you define the noise level?

(c) At what value of S/N

(i) would you believe that you had detected a trace impurity?

(ii) would you be sure you had detected a trace impurity?

(iii) would you be willing to report how much of the impurity was present?

8 | Oestrogens promote growth in cattle, but are banned because of concerns over their entering the food chain. Farmers have been brought to court on suspicion of using these illegal drugs. In such cases an analyst may be required to state the probability of obtaining a false positive result (i.e. detecting trace quantities of illegal oestrogen in meat where none were present). In a particular case where oestrogen was detected the analyst claimed that the chance of obtaining a false positive was 1 in 5000.

(a) Which of the following statements do you agree with?

(i) The analytical evidence establishes guilt beyond reasonable doubt.

(ii) The analytical evidence establishes guilt on the balance of probability.

(iii) On the basis of the analytical evidence alone, there is insufficient evidence to prosecute.

(b) If you were advising one of the lawyers in this case, what points would you make to

(i) the case for the prosecution?

(ii) the case for the defence?

(c) Do you think it would be reasonable to use the same criteria in a case which involved

   (i) a civil case in which a butcher was suing the farmer because of having been supplied with unsaleable meat due to the oestrogen contamination?

   (ii) banning by an athletic association for illegal drug use?

   (iii) prosecution for murder?

ANALYSIS

Making a judgement about the suitability of an analytical procedure involves knowing something about the accuracy and precision of the procedure (see previous sub-section). In addition, it is necessary to take into account many other factors such as suitability for a particular environment, cost, availability of equipment, sensitivity, and so on.

The exercises in this sub-section provide examples which require you to think critically about the features of different analytical problems which help you make a judgement about what makes an appropriate procedure.

1 | An analytical method must be capable of providing a reliable result. In addition, it must be suitable for use in a particular context and environment. For example, it is inappropriate to dissolve a valuable silver object in nitric acid in order to determine the purity of the silver.

You do not need to have a detailed knowledge of the possible analytical procedures in order to recognise the general nature of the constraints which are imposed by specific problems.

(a) You are asked to carry out an *in-situ* analysis of trace contaminants in the Antarctic. How does being in the Antarctic affect the choice of instrumentation you will take?

(b) What principles would you apply in deciding how best to analyse a supposedly valuable painting to check whether it is a forgery?

(c) How would you determine the amount of carbon dioxide in a sample of lemonade in a plastic bottle?

(d) How would you check the composition of evolved gases from an erupting volcano?

(e) In planning a procedure to check the moisture content on-line of paper being produced in a factory at the rate of 15 m s^{-1}, what problems do you foresee and what characteristics of the method are essential?

2 An analytical procedure involves a number of steps which may include:

* Taking a sample of the material to be analysed.
* Dissolving or extracting or otherwise processing the sample.
* Making a measurement with an instrument, which may have some deficiencies.
* Processing the data.
* Reporting the result.

Which of these steps would you expect to be sources of significant errors if you were determining

(a) the percentage of iron in a batch of steel?

(b) the concentration of mercury in fish caught from a trawler?

(c) the fibre content of a sample of a cereal?

(d) the opium content of poppies?

3 The absorbance scale on a spectrometer is linear with respect to concentration, but logarithmic in respect of the radiation detected. Thus a solution of absorbance 1 A.U. transmits 10% of the incident radiation, one of 2 A.U. transmits 1% and so on. Some high precision spectrometers now have scales recording up to 6 A.U.

(a) What proportion of the incident radiation is transmitted by a sample with an absorption of 6 A.U.?

(b) What are the problems and likely sources of error associated with the accurate measurement of an absorbance of 5 or 6 A.U.?

4 The concentration of bromide can be determined by measuring the amount of a standard bromate solution needed to oxidise bromide to bromine. Vogel, *Quantitative Analysis* (3rd edition), Longman, 1962, at page 388 states 'The exact end-point is not easy to detect, and it is better to add a slight excess of standard bromate solution . . . then add 10 ml of 10% potassium iodide solution and titrate the liberated iodine . . .'

NOTHING SUCCEEDS LIKE EXCESS . . . .

This procedure is an example of back-titration. The advantage is that its end point is determined by the disappearance of the starch-iodine colour, whereas the end point of the direct titration is determined by an indistinct colour change of an acid-base indicator.

(a) How much would you consider to be a 'slight excess' of the standard bromate solution?

(b) Compared with the direct titration, does this procedure introduce any additional error?

5 The following techniques are often used to determine the identity and purity of substances.

- NMR spectroscopy
- Mass spectrometry
- GC
- HPLC
- Elemental analysis
- Microscopy
- Infrared spectroscopy
- Atomic absorption spectroscopy

Which of these techniques would you use if you wanted to determine the following

(a) the purity of a sample of hexachlorobenzene.

(b) the purity and identity of a sample of a hydrocarbon $C_{30}H_{62}$.

(c) whether a sample claimed to be coconut charcoal really had been made from coconut husks.

(d) the amount of lead in soil.

(e) the composition of a polymer blend.

(f) the identity of the fibres in a sample of cloth.

(g) the nature of the pigments in a sample of paint.

6 **(a)** We often, or perhaps nearly always, use surrogate measures. For example, when we determine the temperature with a mercury-in-glass thermometer we are actually measuring the length of a thread of mercury. What are we measuring

   (i) when we run a Fourier transform infrared spectrum?

   (ii) when we determine the pH of an aqueous solution by means of a glass electrode?

   (iii) when we carry out an acid-base titration using an indicator?

   (b) Why can we be confident that these things do actually measure what we think they are measuring?

7 In order to have confidence in an analytical result, you need to calibrate your procedure with some appropriate standard. What characteristics would you want in a calibration standard for each of the following procedures?

   (a) Measuring the pH of an aqueous solution.

   (b) Determining the λ_{MAX} of a compound in the uv/vis region of the spectrum.

   (c) Using GC to determine the composition of a mixture of fatty acids obtained by hydrolysis of a sample of triglyceride. (It is necessary to make methyl esters of the fatty acids to increase their volatility.)

   (d) Determining the percentage by weight of Al and Mg (by atomic absorption spectroscopy) and of Si (by ICP-MS) in a sample of mineral.

EQUILIBRIA

This sub-section has many parallels with sub-section 4.3 (Reactions). This is bound to be the case since all reactions have an equilibrium constant. Not surprisingly, therefore, some of the exercises in this sub-section are concerned with making judgements about the meaning of technical words in different contexts. Other exercises ask you to make judgements about procedures and about the likely consequences of carrying out some of these procedures. It is useful to be able to do this since professional chemists carrying out experiments cannot (by definition) know what will happen, and often need to make predictions. This sub-section also introduces some exercises which involve doing a 'back-of-the-envelope calculation'. For some people the results of these calculations may come as a surprise, and may help you to understand more fully the principles of an equilibrium.

The next sub-section builds on the idea of back-of-the-envelope calculations.

1 **(a)** Water dissociates according to the expression

$$2H_2O \rightleftharpoons H_3O^+ + OH^-$$

Does the proportion of dissociated molecules change with

(i) temperature?
(ii) pH?
(iii) dilution of water into, for example, propanone (acetone)?

(b) Hydrochloric acid is usually regarded as being a strong acid and therefore completely dissociated in dilute aqueous solution.

(i) Perchloric acid is a stronger acid than hydrochloric acid; can it be more dissociated than HCl in aqueous solution?
(ii) For what sorts of reasons might it be useful to use an acid which is stronger than HCl?

(c) Calcium carbonate can dissociate as shown:

$$CaCO_3(s) \rightleftharpoons CaO(s) + CO_2(g)$$

The equilibrium constant is 0.168 at 800 °C.
Could you convert $CaCO_3$ into CaO in 100% yield at 800 °C?

(d) Adenosine triphosphate (ATP) plays a key role in energy transfer reactions in living organisms. If A stands for adenosine, the hydrolysis of ATP at body pH can be represented as:

$$A-O-\overset{O}{\underset{O^-}{P}}-O-\overset{O}{\underset{O^-}{P}}-O-\overset{O}{\underset{O^-}{P}}-O^- + H_2O \rightleftharpoons A-O-\overset{O}{\underset{O^-}{P}}-O-\overset{O}{\underset{O^-}{P}}-O^- + O-\overset{O}{\underset{O^-}{P}}-O^- + H^+$$

$\Delta G^\ominus$ is about +7 kJ mol^{-1}.
Would this reaction go to completion

(i) under standard conditions?
(ii) at pH 7.0?

(e) Is there a 'rule of thumb' which you can use to decide when an equilibrium constant is big enough for you to regard the reaction as one which goes to completion?

ITS MY STANDARD
RULE OF THUMB'

2 Many acids dissociate according to the following equation:

$$HA \rightleftharpoons H^+ + A^-$$

The dissociation constant (K_a) is defined as $\dfrac{[H^+][A^-]}{[HA]}$

pK_a for the dissociation is defined as $-\log K_a$

Therefore $pK_a = pH - \log\dfrac{[A^-]}{[HA]} = pH + \log\dfrac{[HA]}{[A^-]}$

(a) When the acid is half dissociated, what can you say about the value of pH?

(b) At what value of $(pH - pK_a)$ would you judge the acid to be:

   (i) 90% ionised?

   (ii) 99% ionised?

   (iii) completely ionised?

3 (a) Is a $0.1\ \text{mol dm}^{-3}$ solution of sulfuric acid stronger than a $0.1\ \text{mol dm}^{-3}$ solution of hydrochloric acid?

(b) Would you expect the pH of $5\ \text{mol dm}^{-3}\ H_2SO_4$ to be

   (i) lower than -1?

   (ii) exactly -1?

   (iii) higher than -1?

4 Consider a coloured species A which ionises according to the following scheme:

$$HA^+ \rightleftharpoons A + H^+ \rightleftharpoons A^- + 2H^+$$
$$\text{p}K_a = 4 \qquad\qquad \text{p}K_a = 9$$

Suppose you wish to determine the molar absorptivity of the species A.

(a) At what pH would you make the measurement?

(b) What variation in pH would you be prepared to tolerate?

(c) What would you use as a buffer, given the list below?

Glycine (amino ethanoic acid)	$\text{p}K_a$ 2.35, 9.8
Acetic acid	$\text{p}K_a$ 4.7
Benzoic acid	$\text{p}K_a$ 4.2
Phosphate	$\text{p}K_a$ 2.1, 7.2, 11.8
Boric acid	$\text{p}K_a$ 9.2

(d) How would your answer be affected

(i) if A could not be protonated?

(ii) if A could not lose a proton?

(iii) if the two pK values were 1 and 6?

5 There is a report in the literature of the sodium salt of an organic acid being recrystallised from 1 mol dm^{-3} H_2SO_4.

(a) In the recrystallising solution, what ions do you expect to be present in significant concentrations?

(b) Given that it was the salt and not the acid which crystallised out, what can you say about the value of the pK of the organic acid?

6 The solubility products for strontium chromate and barium chromate are respectively 3.6×10^{-5} and 5.0×10^{-10} mol dm^{-3}.

(a) At what ratio of Sr:Ba in aqueous solution would you expect both salts to precipitate?

(b) If you had a mixture of Ba^{2+} and Sr^{2+} in aqueous solution, what factors would determine whether or not you regarded titration with potassium chromate as a suitable method for preparing

(i) pure $BaCrO_4$?

(ii) pure $SrCrO_4$?

7 | K_p for the dissociation of N_2O_4 to NO_2 is 0.115 atm at 298 K. Suppose you have two identical samples of N_2O_4 which you maintain at 298 K:

(a) If you compress one sample to one tenth of its original volume will the ratio of N_2O_4:NO_2

    (i) go up?

    (ii) stay the same?

    (iii) go down?

(b) If you add helium to the second sample to increase the pressure to ten times the original value, will the ratio of N_2O_4:NO_2 compared with the ratio in the compressed sample

    (i) be higher?

    (ii) be the same?

    (iii) be lower?

8 | Copper(II) ions complex with ammonia to give a highly coloured product. The stability constant for the complex is 1.4×10^{13}.

$$Cu^{2+} + 4NH_3 \rightleftharpoons [Cu(NH_3)_4]^{2+}$$

Suppose you make a solution of a pure salt of the copper-ammonium complex.

    If you dilute the solution ten times:

(a) Will the ratio of Cu^{2+}:$[Cu(NH_3)_4]^{2+}$

    (i) go up?

    (ii) go down?

    (iii) stay the same?

(b) Will the colour of the solution

    (i) decrease to less than one tenth?

    (ii) decrease by one tenth?

    (iii) decrease by less than one tenth?

9 | One way to determine the concentration of Ni^{2+} is to titrate it with the complexing agent EDTA. Murexide is used as an indicator; this is a blueish substance which forms a yellow coloured complex with Ni^{2+}. The end point of the titration is reached when the colour changes from the yellow of the murexide complex to blueish violet showing that free murexide is present.

The stability constant for the EDTA–Ni^{2+} complex is 4×10^{18} mol dm^{-3}. Typically, the concentration of Ni^{2+} in the solution to be titrated will be about 2×10^{-2} mol dm^{-3}, and murexide indicator will be added to give a concentration of about 10^{-5} mol dm^{-3}.

(a) **(i)** If you add a stoichiometric amount of EDTA, what will be the concentration of free Ni^{2+}?

   (ii) Would the colour have changed from yellow the blue?

   (iii) Is it likely that the colour change would occur too soon? (That is, before a stoichiometric amount of EDTA has been added?)

(b) **(i)** What concentration of EDTA would be required to reduce the concentration of free Ni^{2+} to effectively zero?

   (ii) Could this be done?

ESTIMATIONS

The most famous estimation question is that from a Cambridge University entrance examination question asking candidates how many molecules of Caesar's last breath they might have just inhaled. The questions below are similarly incapable of an exact answer and require estimations based on common sense to produce a rational and sometimes unexpected answer. This usually involves carrying out 'back-of-the-envelope' calculations. In real situations a rough answer is often useful or it may be the best you can do. Sometimes the approximation will enable a decision to be made as to whether to begin to think in detail about the problem, or whether the task is clearly impossible.

1 If all the air in the room you are now in were to be condensed to a liquid, what volume would it occupy?

2 One of the purest materials available is silicon for electronics. It can be 99.999999% pure. How many impurity molecules are there in a silicon wafer of this purity?

3 If all the molecules in a sugar cube were laid end-to-end, how far would they stretch?

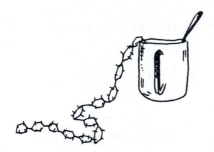

4 Traps for attracting Gypsy moths are commonly baited with a few nanograms (10^{-9} g) of the Gypsy moth pheromone (sex attractant) called disparlure. Such traps are effective in attracting the moths over a distance of at least a kilometre over a period of a few days.

(a) Suggest an approximate concentration of pheromone molecules in the air at a distance of 1 km from the source.

(b) How few molecules do you suppose need to interact with a pheromone receptor to trigger a response from the moth?

(c) What volume of air might a moth pass through when flying around for one minute?

(d) The antennae are believed to carry the receptor sites. Can you think of any mechanism by which the moths can guide themselves in the direction of the source.

5 We shed our outer skin completely approximately every three weeks.

(a) What mass of skin do we lose during this time?

(b) What does this imply in terms of molecules of amino acid incorporation into the skin per second?

6 The highest vacuum that can be obtained is about 10^{-9} torr (pascals).
How many molecules are there in a vessel of 1 litre capacity subjected to such a vacuum?

7 In a real case, a chemical company required 10 tonnes per year of a specialist product which is manufactured by the hydrolysis of a liquid. It was suggested that it could be made by a continuous process in which the liquid was injected into a stream of water through an ordinary laboratory syringe needle.

(a) Could the liquid be injected into the water at a sufficient rate?

(b) What problems can you identify which might arise with this procedure?

8 (a) How much would a mole of peanuts weigh?

(b) How many moles of people populate the earth?

9 In reflection spectroscopy, samples are often diluted with an inert scatterer such as silica. Ideally, in the absence of sample, all the incident light is scattered (i.e. reflected). Roughly 4% of the light incident on a single clean silica surface in air is reflected.

How thick would a layer consisting of 50 µm silica particles have to be before you considered that all the light would be reflected?

10 The proverbial expression 'looking for a needle in a haystack' is sometimes adopted by scientists trying to detect or identify traces of compounds.

(a) If there is one needle in a haystack, estimate its concentration; give your answer in parts per 10^n on a weight or volume basis.

(b) Suppose you made up a solution with a concentration of 'one needle per haystack' (using the value you have just estimated). What volume of the solution would contain a single molecule of solute?

(c) Is the task of looking for a needle in a haystack comparable with using atomic absorption spectroscopy to detect a metal ion at a concentration below 1 ppb?

11 Imagine that you are in the business of providing an analytical service. Consider any instrumental technique you like.

(a) What would you take into account when setting a realistic price to cover your costs for a single analysis?

(b) How would you justify charging a different price if you were providing a regular service for the customer?

12 The pH of the liquid inside the cells of most organisms is about 7.

(a) How many protons would you expect to find in the fluid inside the cell of a typical small bacterium (volume about 2×10^{-12} dm^3), or a mycoplasm (volume about 10^{-15} dm^3)?

(b) Do you suppose the protons would be evenly distributed through the volume?

(c) Do you suppose the number of protons is constant?

(d) Does a pH of 7 have any meaning in the context of a bacterial cell or a mycoplasm?

REFERENCE TRAILS

The chemical literature provides us with an enormous wealth of information. It is so vast that it is impossible to read in detail everything produced even in a specialised area. Tracking down a particular piece of information is not always a straightforward task. It often involves selective reading of complex articles in order to locate items of significance and interest followed by careful reading of these selected items.

An inherent feature of the scientific literature is that the information required may not be located in the immediate paper but in an earlier one referenced in the paper.

The exercises in this section are designed to provide experience of finding information in the primary literature. In each case you are provided with a reference to a paper. This provides a starting point from which to answer the questions or complete the tasks. Some answers will be found in the reference given. To obtain other answers you will be able to use earlier papers referred to by the author to track down the information that you need. In a few cases, an earlier reference may lead you to an even earlier one. Real reference trails often prove unfruitful, especially where general information is being sought. In these examples we have tried to make the tasks specific so that an answer can be located. The answer to Exercise 1 is given here. Answers to all other odd numbered exercises can be found in Section 6.

1 | A paper by Woodward and Schwartz entitled 'In Situ Observation of Self-Assembled Monolayer Growth' (*J. Am. Chem. Soc.*, 1996, **118**, 7861) presents the first *in situ* images of SAM growth proving that molecules aggregate into dense two-dimensional islands when they come out of solution.

(a) What are SAMs?
(b) SAMs of dialkyl ammonium salt monolayers on mica have previously been studied. What method was used to obtain the images? What did these images show?
(c) According to Woodward and Schwartz, what formation process do the SAMs demonstrate?
(d) What are the factors which affect the extent of growth of monolayer islands?

GETTING STARTED

(a) Self assembled monolayers are monolayers of organic molecules that form spontaneously on a solid substrate by adsorption from solution.

(b) Images were collected using LFM, lateral force microscopy, and showed a close packed structure with homogeneity extending to at least a few micrometres. (*Langmuir*, 1994, **10**, 2241).

(c) Formation proceeds by nucleation growth and coalescence of submono-layer islands.

(d) The monolayer island growth is proportional to concentration $\times$ time$^{1/2}$.

2 A paper by Kamounah and Christensen reports 'A New Preparation of Iodoferrocene' via ferroceneboronic acid (*J. Chem. Res.* (S), 1997, 150).

(a) Briefly describe the method used by the authors to produce iodoferrocenes.

(b) Find a method used for the synthesis of iodoferrocene which involves lithiation of the ferrocene.

(c) Give some applications of ferrocenes described by the authors of this paper.

(d) Draw the structure of iodoferrocene.

3 Supercritical fluids are already used in the food industry and could provide an environmentally friendly alternative to solvents in a variety of industrial sectors. Many of these possibilities are discussed in an article entitled 'New Applications of Supercritical Fluids' by Brennecke (*Chem. Ind.*, 1996, 831).

(a) What is a supercritical fluid?

(b) What is RESS?

(c) Supercritical fluids are used to produce products with interesting morphologies. Give some examples.

(d) Explain the abbreviation GAS and describe the principles of the process.

(e) Explain the abbreviation PCA and describe the principles of the process.

(f) What makes supercritical fluid solvents better than organic solvents for many reactions?

(g) What are the values of critical pressure and temperature for CO_2.

(h) Give some examples of applications for supercritical CO_2.

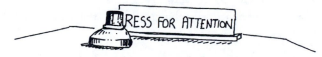
RESS FOR ATTENTION

4 In *J. Chem. Soc., Faraday Trans.*, 1993, **89**, 1945, Kirby, Mallion and Pollak discuss the possible existence of 'toroidal polyhexes'.

  (a) What is a polyhex?

  (b) What is a toroidal polyhex? Draw a diagram of a possible toroidal polyhex.

  (c) Is there any evidence to suggest that any toroidal polyhexes exist?

  (d) Do the authors believe that toroidal polyhex structures could be formed? What would be the factor which most affects their stability?

  (e) What well known class of compounds are polyhexes related to?

  (f) Give the author, title and reference for one other paper which discusses theoretical criteria for the existence of three-dimensional hollow carbon cages.

5 A paper by Galbács and Csányi reports on the alkali induced decomposition of hydrogen peroxide (*J. Chem. Soc., Dalton Trans.*, 1983, 2353).

  (a) In these investigations it was vitally important to keep contaminants to a minimum. The water used for the production of high purity alkali solutions was purified in several stages. Describe how the authors did this.

  (b) In alkaline solutions, what range of pH values was found to result in the maximum rate of decomposition?

  (c) Numerous observations have shown that solid surfaces of different compositions strongly influence the rate of decomposition of hydrogen peroxide. Give some examples of the surfaces which have been investigated.

  (d) Metal ions have been found to catalyse the decomposition process. What methods did the authors use to remove metal ions?

6 Todd (*J. Chem. Ed.*, 1994, **71**, 440) describes the unexpected results of interrupting the distillation of the products of dehydration of 2-methylcyclohexanol.

  (a) Briefly describe the unexpected results. What explanation was given for these results by the authors?

  (b) Use the references in this paper to find more details of the experiment.

(c) Write a detailed experimental procedure for the experiment, including details of apparatus. Your account should be suitable for inclusion in an undergraduate laboratory manual.

7 The paper 'Characterisation of Cr/silica catalysts, Part 2' by Ellison and Overton (*J. Chem. Soc., Faraday Trans.*, 1993, **89**, 4393) describes the preparation and analysis of a series of Cr/silica catalysts and their subsequent investigation by temperature programmed reduction.

(a) How did the authors determine the chromium content of the catalysts?

(b) Sketch a block diagram of a temperature programmed reduction (TPR) instrument.

(c) The authors used CuO as a standard when initially setting up their instrument. What is the literature value for the TPR 'temperature of peak maximum' for a CuO standard.

8 Norrild and Eggert (*J. Chem. Soc., Perkin Trans. II*, 1996, 2583) used boronic acids as potential sensors for carbohydrates.

(a) On what reaction does the sensor mechanism depend?

(b) Find and outline the preparation of *p*-tolylboronic acid, a key compound used in the paper. Draw its structure.

9 A paper by Klapötke, McIntyre and Schulz (*J. Chem. Soc., Dalton Trans.*, 1996, 3237) describes the use of nitryl tetrafluoroborate as a nitrating agent. The authors mention the nitration of adamantane with NO_2^+ salts and with nitryl tetrafluoroborate.

(a) Give the structure of adamantane.

(b) What was the reported yield when adamantane was nitrated with NO_2^+ salts?

(c) What yield was obtained when adamantane was nitrated with nitryl tetrafluoroborate?

10 Komiyama, Sumaoka, Yonezawa, Matsumoto and Yashiro (*J. Chem. Soc., Perkin Trans. II*, 1997, 75) used polyamine cobalt complexes for the catalysis of the hydrolysis of adenosine-3',5'-cyclic monophosphate.

(a) Draw the structures of the following ligands:

- 1,4,7,10-tetraazacyclododecane
- 7(R),14(R)-5,5,7,12,12,14-hexamethyl-1,4,8,11-tetraazacyclotetradecane

(b) Outline a method for the synthesis of tetra-methylethylene diamine.

11 In *J. Chem. Soc., Dalton Trans.*, 1995, 3825, Gareh *et al.* describe a novel method for the preparation of solid compounds.

(a) What is novel about this method of preparation?
(b) Why is it necessary to use a novel method of this kind to prepare metal sulfides from a sulfur source such as Y_2S_3?
(c) What terminology is used to describe the type of reaction in which Y_2O_3 is prepared from Y_2S_3?

12 Beeby, Parker and Williams (*J. Chem. Soc., Perkin Trans II*, 1996, 1565) describe photochemical investigations of ligands incorporating naphthyl chromophores. One of these was

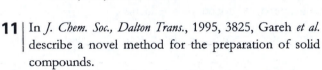
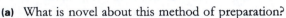

(a) Where is the peak of its fluorescence emission spectrum? How is the intensity affected by the presence of different metal ions?
(b) These authors used a molybdenum tricarbonyl to selectively protect one of the nitrogens. Why might they have considered using a chromium complex?

13 The hydrolysis of 4-nitrophenyl esters of pyridine carboxylic acids were investigated by Colthurst, Nanni and Williams (*J. Chem. Soc., Perkin Trans. II*, 1996, 2285).

(a) Describe how you would synthesise 4-nitrophenyl picolinate and 4-nitrophenyl isonicotinate.

(b) Why did the authors choose this method?

14 Ley and Simpkins report that the anion from phenyl-sulfonyltrimethylsilylmethane reacts in 1,2-dimethoxy-ethane at $-78\,^\circ$C with both aldehydes and ketones to give the corresponding alkenyl phenyl sulfones in good yield (*J. Chem. Soc. Chem. Comm.*, 1983, 1281).

(a) Find an equation illustrating Peterson's phosphorus olefination reaction and summarise the essential features of the reaction.

(b) Ley and Simpkins used a version of the Peterson re-action to convert various carbonyl compounds to vinyl sulfones. Give the reaction scheme for their reaction.

(c) What type of compounds were prepared by a new method during the total synthesis of ajugarin I, the poly-oxygenated diterpene insect antifeedant?

(d) What is the structure of ajugarin I?

15 A paper by Gillespie and Liang entitled 'Superacid Solutions in Hydrogen Fluoride' (*J. Am. Chem. Soc.*, 1988, **110**, 6053) discusses studies of the acidity of various solutions of inorganic compounds in HF using nitroaromatic compounds as indicators.

(a) Clifford and co-workers previously determined the acid strengths of metal fluorides in HF on the basis of the ability of the solutions to dissolve CoF_3. What order of acid strength did they obtain and which other metal ions were found to dissolve in HF solutions of SbF_5?

(b) How did McCauley *et al.* evaluate the acid strength of metal fluorides in HF. What order of acid strength did they obtain and to what was the acid strength attributed?

(c) During their investigations Gillespie and Liang obtained the order of strength of the pentafluorides. State their findings.

(d) Give the equations for the reactions of SbF_5 and AsF_5 with HF.

(e) Gillespie and Moss report the identification of polynuclear species such as $Sb_2F_{11}^-$ when studying SbF_5 in HF. Give the equation for the equilibrium between the monomeric and the dimeric species. What other system did they study?

16 Assume that you wish to do some work similar to that described by Naud in his paper 'Thermal decomposition of aryl nitramines' (*J. Chem. Soc., Perkin Trans. II*, 1996, 1321), and need to obtain the rate constants in order to determine the Arrhenius parameters for the decomposition of nitramines.

Consider the compound below.

(a) Give an outline of the method you might use to prepare this compound.

(b) What sort of rate constant for the decomposition reaction would you expect?

(c) Why would you carry out the decomposition reactions in toluene or benzene?

(d) Why would you add an excess of organic base to the reaction?

17 In a paper by Ikeyama, Kabuto and Sato (*J. Phys. Chem.*, 1996, **100**, 19289), the Huang–Minlon method was used for the preparation of 5H-dibenzo*[a,d]*cycloheptene.

(a) What is the Huang–Minlon method and what is it used for?

(b) Write a description of the preparation of 5H-dibenzo*[a,d]*cycloheptene suitable for use in an undergraduate laboratory manual.

18 In the paper 'Radical cations from nitrone spin-traps: reaction with water to give OH adducts', Bhattacharjee, Khan, Chandra and Symons discuss the use of nitrone spin-traps to detect OH radicals in biological systems by liquid phase EPR spectroscopy (*J. Chem. Soc., Perkin Trans. II*, 1996, 2631).

(a) What is spin-trapping and why is it particularly useful for detecting radicals such as OH and O_2^-?

(b) Which three compounds were used as spin-traps in this study?

(c) Find the EPR spectrum of a normal OH radical and sketch it.

19 The paper 'SnO_2-V_2O_5 based catalysts' by Cavani, Trifior, Bartolini, Ghisletti, Nalli and Santucci (*J. Chem. Soc., Faraday Trans.*, 1996, **92**, 4321) describes the preparation, analysis and reactions of some catalysts.

How did the authors determine the vanadium content of the samples?

20 Chromatographic isomer separation and circular dichroism of rhodanines was discussed by Rang, Isaksson and Sandström (*J. Chem. Soc., Perkin Trans. II*, 1996, 1493).

(a) Draw an outline diagram of the equipment used to determine the temperature dependence of the CD spectrum.

(b) What was the stationary phase used for the separation?

(c) Why was special apparatus used?

21 In the paper by Rondeau, Raoult, Tallec, Sinbandhit, Toupet, Imberty and Pradere (*J. Chem. Soc., Perkin Trans. II*, 1996, 2623) stereoselective electrochemical reduction of thiapyrans was discussed. For comparison, chemical reduction with zinc and acetic acid was also carried out, using methods previously developed for the selective reduction of α,β-unsaturated ketones.

(a) Give a general procedure for the chemical reduction.

(b) What method was used to determine the ratio of diastereoisomeric products from the electrochemical reaction?

22 In 1969, Timms reported the first preparative co-condensation reactions of metal vapours with organic and inorganic substrates. If metal atom reactions are not utilised primarily for the formation of new products, but for the synthesis of highly reactive intermediates, a new synthetic strategy may be developed. This is the subject of the paper 'Highly Reactive Intermediates From The Co-condensation Reactions of Ion, Cobalt & Nickel Vapour with Arenes' (Zenneck, *Angew. Chem. Int. Ed. Engl.*, 1990, **29**, 126).

(a) Give a brief description of the method used by Timms to prepare some of the metal complexes.

(b) To what type of ligands is the above method limited?

(c) Which properties of the boron heterocycles do the authors think are interesting?

(d) Draw the structure of the red compound which is formed as the product of the reaction of $Co(arene)_2$ with acetonitriles rather than alkynes.

(e) What three methods were used in the synthesis of bis(2,2'-bipyridyl) cobalt?

23 In the paper 'Thermal reactions of $Mo(CO)_6$ on metal oxide surfaces' (*J. Chem. Soc., Dalton Trans.*, 1995, 3753) Ang, Chan, Chuah, Jaenicke and Neo describe their investigation of the decomposition of $Mo(CO)_6$ and their determination of the dissociation energy of the metal-carbonyl bond.

(a) The activation energy for the decomposition process is quoted as 80–100 kJ mol^{-1}. What method was used to determine this activation energy?

(b) Find a value for the bond energy for the rupture of the first Mo–CO bond. How was this bond energy determined?

24 A paper by Nyokongin (*J. Chem. Soc., Dalton Trans.*, 1993, 3601), describes the investigation of the kinetics and equilibria of the interaction of CN with the Fe(II) complex of phthalocyanine.

(a) Why are metal phthalocyanine complexes commercially important?

(b) The Fe(II) phthalocyanine species is particularly important because of the interest in possible antidotes to cyanide poisoning. What species in biology is phthalocyanine analogous to?

(c) Why is direct comparison of the two species referred to in (b) unreliable?

(d) What are the equations for the reactions of cyanide with [Fe(II) (pc)] in a coordinating solvent?

(e) What are the equilibrium constants for the above reactions in dimethyl sulfoxide?

(f) What would the overall equilibrium constant be for the reaction in dimethyl sulfoxide?

(g) The authors give details for the preparation of $K_2[Fe(pc)(CN)_2]$ which was used in their investigation. Rewrite this preparation as if it was going to be used as an undergraduate laboratory exercise.

25 In their paper in *J. Chem. Soc., Faraday Trans.*, 1993, **89**, 95, Maity, Palit, Mohan and Mittal describe their investigation into the stability of fullerenes to radiation.

(a) When, where and by whom was the first paper describing fullerenes published. By what method were the first fullerenes prepared?

(b) What do previous studies suggest about the stability of fullerenes to light?

(c) A paper was published on this same subject shortly before the paper by Maity *et al.* Which types of radiolysis are discussed in each paper?

(d) Give the experimental conditions described in each paper for the common method of radiolysis.

(e) What were the conclusions of the two sets of workers? What do they suggest as the electronic excited state?

(f) Is C_{60} or C_{70} more stable to radiolysis? Is the stability affected by the solvent used?

26 | A paper by Garrou (*Chemical Reviews*, 1985, **85**, 171) has the title 'Transition-metal mediated phosphorus-carbon bond cleavage and its relevance to homogeneous catalyst deactivation'.

(a) Nyholm found that decomposition of $Os_3(CO)_{10}(PPh_3)_2$ yielded several products. How many products were formed and what were their chemical formulae?

(b) Which two research groups simultaneously discovered that refluxing $IrHCO(PPh_3)$ in a high boiling solvent resulted in the separation of purple diamagnetic crystals? Give the structure of these crystals and a brief outline of the method used for their preparation.

(c) *Meta-* and *para-*substituted aryl-nickel complexes gave three unexpected products. What were they?

(d) Among the many catalyst systems patented for the homologation reaction of methanol to ethanol and higher alcohols, how is iodide thought to catalyse the reaction?

27 | In the paper 'Dinuclear palladium(II) complexes with isocyanide and *N*-donor bidentate ligands' (*J. Chem. Soc., Dalton Trans.*, 1996, 3059), Tamasi Ukaji and Yama Moto used the compound $[Pd_2Cl_2(2,4,6-Me_3C_6H_2NC)_4]$ as one of their starting materials in the synthesis of new complexes.

Describe briefly how you would prepare $[Pd_2Cl_2(2,4,6-Me_3C_6H_2NC)_4]$.

28 A paper by Chambers *et al.* (*J. Chem. Soc., Perkin Trans. I*, 1996, 1659) describes a new method for halogenating aromatic compounds. It also refers to the earlier use of *N*-iodosuccinimide as a useful iodinating reagent.

Find a paper describing the use of this reagent to prepare *trans*-1,2-iodo-carboxylates from alkenes.

If you wanted to use this method

(a) in what solvent would you mix the reactants?

(b) what reactants would you mix?

(c) in what ratio would you mix the reactants?

COMMENTARIES

3 | **Our choice is D.**

We know that thixotropic substances exist, and it is possible that the repeated movement of the phial during the ceremony could be sufficient to cause a change in state.

A has been supported on the grounds that a 'miracle' (or supernatural event) can be defined as any event which defies explanation. We accept this argument, but suggest that good science requires that a testable hypothesis be proposed which may provide an explanation.

B. We are unaware of reversible photochemical processes which involve changes between the liquid and the solid state. This does not mean that they do not or could not exist. But it seems to us less plausible than D. B is only plausible if it is assumed that whoever demonstrates the liquefaction process shields the phial from light for varying periods.

C. We find this an implausible hypothesis because periodic growth, even if this could be instigated, is not a reversible process; we cannot imagine how the change from solid to liquid could depend on whether or not growth was occurring.

E. In order for a hygroscopic substance to pick up moisture from the atmosphere it must obviously be in an open vessel. The fact that the vessel is turned upside down suggests that it must be a closed one.

5 | **Our choice is C.**

The argument is that it is not worth trying to synthesise the ester from acid and alcohol *because the purification involves separating ester from unreacted alcohol.* C is the only option offered which addresses the flaw in this argument.

A and D are both true, and offer ways of avoiding the problem. However, they do not deal with the flawed statement in the second sentence.

Catalysts affect rates of reaction, but cannot change an equilibrium constant. It follows that B cannot be true unless it is interpreted to mean that suitable catalysts are specific for esterification reactions. If this interpretation is used, B is irrelevant to the argument.

7 | **Our choice is B.**

The first sentence in this passage is crucial. If you do not believe it to be true, there is no reason why you should accept the argument. B provides an explanation for the loss of hydrogen and helium from the atmosphere, and thus puts the argument on a strong foundation.

A is not true; B explains what the mechanism is. If you are not convinced by B, then you might choose A — and *not* accept the argument.

C is true, but not relevant to the question since the passage states that both hydrogen and helium have been lost from the atmosphere.

D is true — and the amount of hydrogen in the atmosphere is very small.

E is true, but not relevant to the question.

9 | **Our choice is D.**

It is important to recognise that this passage is drawing conclusions about the *reasons* for developing lead-free petrol. D is the only sentence of those given which actually addresses this argument. In order to recognise D as a *flaw* in the argument as written, you need to *know* that D is true: it is.

A is true, and indeed catalytic converters which become poisoned by lead are not thrown away — whether they are based on platinum or (for example) palladium; the valuable metals are purified and recycled. However, if the purification was really cheap, there would be no need to use lead-free petrol in cars fitted with catalytic converters. Thus, A is not a flaw in the argument as stated.

B and C are both true, but neither are relevant to the argument as it is given here.

11 | **Our choice is B.**

Conduction of electricity is movement of electrons. The electrons can move freely within the sheet, but fail to move from one sheet to another. The most reasonable explanation is that the gap is too large.

A is merely a repetition of information given in the passage.

D is true, but states nothing about electrical conduction.

C is not strictly correct; it is true that high conductivity is associated with delocalised electrons, and that electrons in π bonds are delocalised. However, C cannot really be described as a theoretical explanation — though it might be considered an underlying assumption in the argument.

13 | **Our choice is B.**

The argument here is about the consistency of the two analysts; consistency must be assessed by looking at the *reproducibility* of results and this is measured by the standard deviation. This being so, it may seem obvious that the analyst with the lower standard deviation is the more consistent. However, imagine instead that we were dealing with 10 measurements made by the same analyst. If we divided the results arbitrarily into two batches of 5, we would not expect to find that both batches had either the same mean or the same standard deviation. A standard deviation based on a small number of measurements is only an estimate; the true value may be larger or smaller.

A may be true, though the data given here lend no support to the suggestion. Even if there were a systematic difference between the two analysts it would not necessarily show that one was more *consistent* than the other – and it certainly would not show that A was the more consistent.

C and D are both true; but not relevant to the question asked.

15 | **Our choice is C.**

Unless C is true, then absence of a neutral substance in the ether layer gives no information about its existence in the original aqueous solution.

A, D and F are irrelevant since nothing is detected in the ether layer. Only if something were found in the ether layer would it be necessary to eliminate the possibility of it being phenol (or a phenyl ester).

Of course, water and ether are not completely immiscible (E). But they are *substantially* immiscible – otherwise ether extraction could not be used. The phrase 'The layers are separated' could be described as a *statement* that water and ether are immiscible; thus E is not best described as an *assumption*.

17 | **Our choice is D.**

Keto-enol tautomerism means that the keto and the enol form are in equilibrium. Under most conditions, individual molecules interchange rapidly between the two forms. Thus, over a period of time, all molecules will exist in the enol form and can react as such. Provided that at least a small proportion of the substance is in the enol form, then A, B and E are all true, but are not relevant to the question asked here.

C represents a way of showing that the test given is capable of differentiating between ketones and enols. It is important to know that this is true, and a sensible experimentalist would actually do this test in order to shown that *in their hands* the test works. Nevertheless, a competent experimentalist would also recognise that the test is actually incapable of leading to the stated conclusion however it was carried out. Therefore C is not a flaw in the argument.

19 | **Our choice is A.**

A is true; ion exchange chromatography would be a much less powerful technique if it were not. It does not follow that the two compounds x and y *can* be separated by ion exchange chromatography. It may be that only trying it out will provide the information. On the other hand, the fact that there can be no difference in charge may be a good reason for trying other separation techniques since ionic interactions with sulfonic acid groups cannot themselves be contributing to the separation.

B and C are both true, but are not relevant to the argument about whether or not x and y can be separated on an ion exchange resin carrying sulfonic acid groups.

D. In this example, changing the pH is unlikely to change the *relative* affinities of x and y for the sulfonic acid groups on the support material (though it may affect the *absolute* affinities). pH cannot affect the charge on the compounds themselves since they are quaternary ammonium ions. If the strength of interaction with x and y depends *only* on the average value of their positive charge (as stated in the passage), then any effect of pH on the state of ionisation of the sulfonic acid groups will affect both x and y equally.

21 | **Our choice is C.**

C is regarded today as the best evidence of delocalised electrons in aromatic systems. The proton peak is shifted upfield compared with normal protons; this is consistent with them being deshielded. This is interpreted to mean that there is a circle of loosely bound electrons which maintain a ring current.

A offers no *further* support for the concept of delocalisation than is already stated in the passage; it simply provides an example of the statement that 'benzene does not react like an alkene'.

D is concerned with the theoretical rationalisation of aromatic behaviour. It is arguable that a theory which is based on limited observations but then turns out to be widely applicable actually strengthens the arguments which led to the development of the theory. But this is not the way in which D is expressed and so, in our view, it does not strengthen the conclusion of the argument given here.

B is often quoted as evidence in favour of the structure of benzene. In fact, the data collected by X-ray diffraction gives values for the *average* position of atoms within a molecule. With very high resolution studies, which can be made using modern techniques and instrumentation, it is possible to obtain evidence that the relative positions of all the atoms in a molecule are not always constant. However, in 1929, when Kathleen Lonsdale first showed the presence of a hexagonal ring in hexamethyl benzene, the resolution was not good enough to disprove the possibility that the bonds in the ring oscillate rapidly between double and single character.

23 | Our choice is C.

A correlation coefficient of 1.00 signifies that data fit a perfect straight line. One of zero indicates that data are randomly scattered within the x, y axes. Values between 0 and 1 indicate whether the data are more closely correlated (nearer to a straight line) than would be expected if there were no relationship between x, y values. Data obtained from a section of a curve will almost always look more like a straight line than like randomly scattered x, y values: such data almost always give a high correlation coefficient. You can try this by calculating exact x, y pairs from an equation such as $y = ax/(b + x)$. By limiting the range of x values you will find that you can generate sets of x, y pairs which give correlation coefficients of 0.98 or greater. Thus a high correlation coefficient is no guarantee of a linear relationship. Plotting a graph will give you a better idea.

It might help if you systematically follow the change in correlation coefficient as the *range* of x values is increased – but it is bound to be more work than the more reliable step of plotting the graph, even if it is possible to collect more data.

B is no help for the reasons given above.

D is always a useless thing to do unless you have a reason for believing that the data are linearly correlated; if you believe that they are, then this is a way of obtaining values for the slope and intercept.

25 | Our choice is E.

There are two key points in this passage. The first is that when results vary (through experimental error or through genuine differences between samples) there is no known way to give an answer with absolute certainty. The supplier is therefore, at least in theory, taking a risk in providing an assurance that the level of the by-product is below the acceptable maximum. The supplier must decide on the acceptable level of risk.

The second key point is that the error quoted in this passage cannot be meaningfully interpreted since the quoted error is not defined; it could be a standard deviation, a standard error, or some kind of confidence limit (for example 95%). The interpretation of the analytical data depends critically on knowing precisely what the ± means.

We see no reason for concluding A; there are many reasons for specifying low acceptable levels of impurity of which 'hazardous' may be one.

Because the meaning of '±1.5' is not defined (see comment above), it is only possible to draw conclusion B, C, or D by making specific assumptions about its meaning.

27 | Our choice is B.

The conventional way to determine $\Delta H^{\ominus}$ and $\Delta S^{\ominus}$ for any process is to measure the equilibrium constant (and hence $\Delta G^{\ominus}$) at different temperatures. The fact that the equilibrium constant for the partition of pentanol between water and pentanol (i.e. its solubility) decreases as temperature increases (B) is evidence for an unfavourable entropy term – as stated in the first sentence in the passage. This does not *prove* that the water molecules reorientate themselves, but it is at least consistent with the suggestion; nowadays it is generally accepted as being the most plausible rationalisation of the observation.

A is true for many substances, but not for pentanol and other non-polar molecules; it is not relevant to the argument here.

C is true and provides (at least part of) an explanation for compounds being more soluble in hot water.

D and E are important general concepts, but do not help to relate the solubility of pentanol to entropy factors.

29 | **Our choice is C.**

The evidence given in the passage cannot give any direct information about electron density, although the inference seems entirely reasonable. In order to confirm the conclusion based on inference, more direct evidence of high electron density is required. Providing that the intermediate is stable enough (as here) this can be obtained from ^{13}C NMR studies since electron density is one of the determinants of chemical shifts. Here ^{13}C NMR studies indicate that carbon 4 is indeed the site of highest electron density, and thus explains why attack by the ammonium ion occurs here.

A and D are alternative expressions of the need to examine the relative stability of cyclohexa-1,3-diene and cyclohexa-1,4-diene. This is stated in the passage to be already established. The case for selecting one of these must rest on an unwillingness to accept that the 'thermodynamic evidence' is convincing.

There is no way in which measuring the rate of reaction (B) can give direct information about the relative electron densities on the carbon atoms, though kinetic studies are often important in deducing *mechanisms* from which relative electron densities can be inferred (as is done in this passage).

CONSTRUCTING AN ARGUMENT

3 | **Our preferred sequence is B, A, this is illustrated by the fact that C.**

A carbon atom is described as chiral when it is covalently bonded to four different substituents. Compounds containing a single chiral carbon atom have two enantiomers each of which rotates plane-polarised light. This is illustrated by the fact that each enantiomer of lactic acid (I) rotates plane-polarised light.

Many chemical arguments include some kind of observation (such as sentence C in this exercise). The observation may seem to be the natural starting point from which logical deductions are made. However, in this exercise sentences A and B are not structured as logical deductions derived from C; when read together they provide a brief summary of the meaning of chiral. This is why we suggest that, in this case, the *observation* appropriately forms the concluding sentence (compare this with the next exercise in this section, Exercise 4).

Notice our choice of words for linking A with C. The sequence 'B, A, *therefore* C' is logically perfectly acceptable. We prefer our choice of linking words because the observations on lactic acid preceded the understanding of chirality. For this reason the process of science is better represented by our linking phrase than by the simpler 'therefore'. See Exercise 5 and several others.

5 | **Our preferred sequence is C, B, this illustrates that A.**

$HClO_4$ has more oxo groups than $HClO_3$. $HClO_4$ is a stronger acid than $HClO_3$. This illustrates that for mononuclear oxoacids, the species with the greater number of oxo groups has the lower pK_a.

This exercise has some parallels with Exercise 3. The sequence 'A, C, therefore B' is perfectly logical. However, we suggest that the observations B and C almost certainly contributed to the establishment of the generalisation or 'rule' A. In this sense we suggest that our preferred sequence better represents the underlying science.

Another possible solution for this exercise involves changing the linking words to something like 'C, B, this leads to the hypothesis that A'. We accept this style of argument. However, in this case, we suggest that it would be undesirable to propose the hypothesis A on the basis of only one pair of compounds.

7 | **Our preferred sequence is A, C, this explains why B.**

Graphite has a layer structure. The layers of graphite are held together by van der Waals forces. This explains why graphite is soft and can be used as a lubricant.

We suggest that the only alternative for a concluding sentence might be C, but our view is that C could not be deduced from the combination of A and B. One might argue that C is actually deduced from knowledge of the layer structure – and specifically from the distance between the layers. But this is not relevant here, since the exercise involves the use of all three sentences in a logical way.

We prefer our linking words 'this explains why' to the simpler 'therefore'. Our preference reflects our belief that B could not be *predicted* from A and C. (Contrast this with Exercise 3.) However, given that B is an observation, A and C together provide a plausible explanation.

We recognise that another possibility is 'A, B, because C'. This use of 'because' leaves B as the conclusion whilst putting the explanatory C at the end of the passage.

(We are aware that this plausible explanation is in fact oversimplified. Completely dry graphite does not act as a lubricant. It therefore seems that this property depends on the presence of water molecules between the layers.)

9 | **Our preferred sequence is A, C, therefore B.**

A solution containing thiocyanate gives a red colour with the Fe^{3+} ion. Solution X contains thiocyanate. Therefore solution X gives a red colour with Fe^{3+}.

Beware of the temptation to give B, A, therefore C. This is another example of false logic (rather more extreme than the possible one in Exercise 1). Substances other than thiocyanate (for example phenols) will give a red colour with Fe^{3+}, and so this is not a definitive test for thiocyanate. In this example, we use the linking word 'therefore' even though it would be more appropriately followed by 'solution X *will give* a red colour with Fe^{3+}'. This illustrates the kind of problem to which we referred in the introduction to Section 1 – namely that we have tried to avoid giving clues to the sequence by the precise wording.

11 | **Our preferred sequence is A, C, this illustrates that B.**

Methanol forms hydrogen bonds and is a liquid. Methane does not form hydrogen bonds and is a gas. This illustrates that the ability to form hydrogen bonds may have an important effect on the physical properties of a compound.

We do not think there is much doubt about the most appropriate sequence here. Other linking words or phrases (such as 'this suggests that' or 'this leads to the hypothesis that' may be preferred. Our view is that (as with Exercise 5) the pair of observations (A and C) are best described as *illustrating* the general principle given in B.

13 | **Our preferred sequence is A, C, therefore B.**

The equilibrium constant for a reaction is defined as the product of the activities of products divided by the product of the activities of reactants, when the system is at equilibrium. When A and B are in equilibrium, their activities are equal. Therefore for the reaction $A \rightleftharpoons B$, the equilibrium constant is 1.

This exercise has some parallels with Exercise 5. The sequence 'A, B, there-fore C' is completely logical. The choice between this and the one we prefer depends on the view one takes of the origin of the sentences B and C.

Our view is that C is likely to be an observation from which B can be concluded, and therefore our preferred sequence is likely to be the best representation of the science (see Exercise 3).

Of course it is possible to calculate the equilibrium constant of a reaction from standard free energy changes of other related reactions. If this were done, C would be a conclusion which properly represented the scientific process.

We also recognise that a useful step in consolidating your under-standing of the concept of equilibria involves calculating activities (as in C) from a given value of an equilibrium constant (as given in B). In our view it is useful to recognise explicitly that the value of this kind of calcula-tion is that it provides opportunities to manipulate data and to strengthen understanding of important concepts; it is important to recognise that such a calculation is an *exercise*. Our knowledge and understanding of chem-istry cannot be increased by the circular process of calculating the activ-ities of constituents (C) from an equilibrium constant (A), which was originally calculated by direct observation of those activities.

15 | **Our preferred sequence is A, B, this illustrates the fact that C.**

Ammonium salts of nitrate, chlorate, and perchlorate are unstable and readily decom-pose when heated. A common feature of the nitrate, chlorate, and perchlorate ions is that they are strong oxidants. Therefore ammonium salts of oxidising anions are unstable.

The sequence 'B, C, therefore A' appears to be equally acceptable from the point of view of logic. However, the logic demands that statement C has been established from observations of the instability of the ammonium salts of oxidising anions (but not of the nitrate, chlorate and perchlorate). Statement C is then used to predict the behaviour of these ammonium salts. For reasons discussed in the commentary on Exercise 5, our view is that the sequence 'A, B, therefore C' is a better repres-entation of the science.

Linking phrases such as 'this leads to the hypothesis that' can properly be defended here, since C could reasonably be seen as an hypothesis based on observation of the behaviour of three common examples of oxidising anions. Contrast this with Exercises 5 and 11 where only a single example is used.

17 | **Our preferred sequence is A, B, therefore C.**

The sum of the covalent radii for P and O is 0.176 nm. In phosphorus pentoxide, P_2O_5, different P–O bonds have bond lengths of 0.160 and 0.143 nm respectively. Therefore P and O are capable of forming a dπ–pπ bond.

The sequence 'A, C, this explains why B' is logically perfectly sound; however, if this sequence is preferable it is important not to use 'therefore' as a linking word, since the formation of a dπ–pπ bond cannot predict exact values for the shortening of the bands.

We prefer C as the conclusion since the two observations (A and B) lead directly to the conclusion that the P–O bonds in P_2O_5, are stronger than simple covalent bonds. The actual conclusion drawn (C) carries an underlying assumption that the nature of a dπ–pπ bond is familiar to the reader.

Using the sequence 'A, B, C', different linking words (such as 'this is evidence that') are perfectly sensible.

19 | **Our preferred sequence is C, B, this is illustrated by the fact that A.**

Octahedral complexes with low spin d^6 electron configurations are inert to ligand substitution. Many Co(III) complexes are octahedral and have low spin d^6 electron configurations. This is illustrated by the fact that many Co(III) complexes are inert to substitution.

Making B the conclusion ('A, C therefore B', or 'C, A therefore B') involves the unjustified assumption that C gives the *only* reason for inertness to ligand substitution.

Statements A and B together provide some support for C, and one can argue that 'A, B, this leads to the hypothesis that C' is logical. However, as with Exercise 11, we suggest that more examples are needed before this hypothesis could reasonably be made.

In our preferred sequence we suggest the linking phrase 'this is illustrated by the fact that'. The use of 'therefore' makes sense in verbal logic. However, as with other examples, the use of 'therefore' implies that observation A would actually be *predicted* from C and B, and obscures the scientific logic that A is one of the observations from which the generalisation C was developed. We reject the linking phrase 'this explains why' on the grounds that B and C do not provide a true explanation.

21 | Our preferred sequence is A, C, therefore B.

Saturated aliphatic esters absorb in the infrared at around 1740 cm⁻¹. Compound A is a saturated aliphatic ester. Therefore compound A absorbs infrared radiation at around 1740 cm⁻¹.

The sequence 'A, B, therefore C' is an example of false logic similar to that in Exercise 9; saturated aliphatic esters are not the only compounds which have an absorption band at around 1740 cm⁻¹. The same objection does not apply to the sequence 'B, A, it is therefore probable that C', since this linking phrase implies that C is an interpretation which must be confirmed by some other observation.

We suggest that it would be unreasonable to draw the conclusion A from observations B and C on a single compound. In contrast, the conclusion B follows logically and directly if C and A are both correct statements.

23 | Our preferred sequence is C, A, this suggests that B.

sp³ hybridisation gives rise to tetrahedral molecules. SiCl₄ is a tetrahedral molecule. This suggests that SiCl₄ is sp³ hybridised.

In our view this sequence demands the use of the linking words 'this suggests that' in preference to 'therefore' since the latter implies that the only possible factor which causes molecules to be tetrahedral is sp³ hybridisation.

'C, B, therefore A' or 'B, C, therefore A' make logical arguments. Our objection to them is that using B to conclude A implies that evidence for B can be obtained independently of A. It is true that, if no other information was available, theoretical considerations would lead to the prediction that SiCl₄ would be sp³ hybridised and therefore would be tetrahedral. Our view is that, 'C, B, therefore A' or 'B, C, therefore A' can only be strongly defended if B is rewritten as 'theory suggests that SiCl₄ is likely to be sp³ hybridised'.

Our objection to C as a conclusion is that it represents another example of formulating a generalisation on the basis of a single example.

25 | Our preferred sequence is A, C, therefore B.

Only compounds with asymmetric electron distribution have a dipole moment. $Ge(CH_3)_4$ is spherically symmetric. Therefore $Ge(CH_3)_4$ does not have a dipole moment.

Our preference relies on a fairly precise understanding of the meaning of symmetry. Sentences A and B together cannot lead to the conclusion that $Ge(CH_3)_4$ is spherically symmetric, since the rule stated in A excludes both centrosymmetric and spherically symmetric molecules from those which have dipole moments. This objection does not arise with the sequence 'A, C, therefore B'.

27 | Our preferred sequence is B, C this explains why A.

BF_3 has six valence electrons. Ammonia is a Lewis base. This explains why BF_3 readily forms an adduct with ammonia.

Our preference over 'C, B, this explains why A' is slight, but our own rules require us to select *one* preferred sequence. We believe that the interest in the final sentence lies with the reaction of BF_3 rather than ammonia. We therefore prefer to start the argument with the statement concerning BF_3.

In this case (in contrast to Exercise 19) we regard the linking phrase 'this explains why' as valid. Both statements C and B are concepts based on theoretical development of experimental observation which, when properly understood, provide a theoretical explanation for A.

The sequence 'B, A, therefore C' is logically defensible. It uses elementary valence theory (B) together with an experimentally verifiable statement (A) to predict a characteristic (C). Our view is that the conclusion C is sufficiently well established that it would not, in fact, be drawn from B and A.

29 | Our preferred sequence is B, C, this illustrates that A.

The (six coordinate) radii of Na^+ and K^+ are 0.102 nm and 0.138 nm respectively. Potassium salts of large anions are generally less soluble than the corresponding sodium *salts. This illustrates that salts containing anions and cations of widely different radii are generally more water soluble than salts containing ions with similar radii.*

The choice of a concluding sentence must be between A and C; the specific values for ionic radii given in B cannot be concluded from A and C, though a passage with a structure like 'A, C, this is consistent with the fact that B' could be defended.

We suggest that B (a very specific statement or observation about sodium and potassium ions), together with C (a rather general statement about sodium and potassium salts) provide an illustration of A; they do not *show* that A is true, hence our use of a linking phrase in preference to 'therefore'.

The sequences 'A, B, therefore C' or 'B, A, therefore C' are logically correct. Our objection is that, historically, C contributed to the generalisation A. In this case, we feel that a linking phrase like 'this explains why' is no help, since A is written as an empirical observation and not as a causal explanation.

Notice that whatever sequence is preferred, there is an underlying assumption that large anions have radii greater than that of potassium; we believe that this is an assumption which is easy to accept.

31 | Our preferred sequence is B, C, therefore A.

The reaction $3H_2 + N_2 \rightleftharpoons 2NH_3$ is exothermic, so le Chatelier's principle indicates that ammonia production would be favoured by low temperature. At room temperature, the rate of reaction between nitrogen and hydrogen is too slow to be commercially useful. Therefore the Haber process used to manufacture ammonia from nitrogen and hydrogen is carried out at a temperature between 400 and 500 °C.

These three sentences can be put together in several different logical sequences. The choice between them (and hence the choice of linking words) must depend on the context which is assumed.

We prefer our sequence because it reflects something of the history of the chemistry of this important reaction. 'B, A, (or A, B) this is because C' would be a perfectly logical way to explain the observation A to someone who knew about the thermodynamics of the reaction (B), but had not thought about the importance of kinetics.

The linking words 'this explains why' could be preferred to 'therefore' on the grounds that C only tells you that the industrial process must take place at a temperature above ambient and does not allow a conclusion about the compromise temperature that will be used.

33 | **Our preferred sequence is A, C, therefore B.**

Many molecules with several conjugated double bonds are coloured. The energy of a photon from the visible region of the spectrum lies between 2×10^{-19} and 5×10^{-19} Joules. Therefore the energy required to excite an electron in molecules with conjugated double bonds, is often between 2×10^{-19} and 5×10^{-19} Joules.

In our view there is little to choose between this sequence and 'C, A, therefore B'; both link a simple observation (A) with a well known physical concept (C) to draw a conclusion (B). We accept that 'B, C, this explains why A' or 'C, B, this explains why A' can be defended by using criteria we have used in previous exercises. But in this case we see no reason for choosing the more complex form of argument which requires the use of a linking *phrase* ('this explains why') rather than 'therefore'.

We reject C as a possible conclusion because B must depend on knowledge of C rather than the other way round.

35 | **Our preferred sequence is C, B, therefore A.**

Each carbon atom in a two-dimensional sheet of graphite is bonded to three others in the plane (bond length 0.140 nm) whereas in diamond the carbon atoms are tetrahedrally arranged and the bond length is 0.154 nm. In graphite, two-dimensional sheets of carbon atoms are separated by a distance of approximately 0.335 nm. Therefore the density of graphite is lower than the density of diamond.

There is an underlying assumption in this argument that close packing of atoms is one of the factors contributing to high density – so that, when two allotropes are compared, the one with the most closely packed atoms will be the densest. We also assume that it is intuitively clear from the bond lengths given that the carbon atoms are more closely packed in diamond than they are in graphite.

In our view 'C, B, therefore A' is a slightly more natural sequence than 'B, C, therefore A' since C introduces the comparison between the two allotropes and this then leads to the conclusion via a single extra piece of information (B).

We can see reasons for preferring the more complex linking phrase 'this explains why' over 'therefore'.

37 | Our preferred sequence is A, C, therefore B.

H_3AsO_4 reacts with iodide in aqueous solution to form an equilibrium mixture with $HAsO_2$ and iodine. Tin(II) chloride reduces iodine to iodide. Therefore when excess tin(II) chloride is added to H_3AsO_4 in the presence of iodide all of the H_3AsO_4 is converted into $HAsO_2$.

In this case we suggest that the argument must start with the primary observation (A).

C does not follow logically from the two observational statements A and B; on the basis of the information given one might equally conclude (for example) that tin(II) chloride directly converts H_3AsO_4 into $HAsO_2$. At best, C is a *plausible* explanation for the two observations A and B and, in order to defend it as a conclusion, it as necessary to make assumptions about the chemistry involved.

No such objection is attached to our preferred sequence, where, if C is known, B must surely follow.

39 | Our preferred sequence is A, B, therefore C or B, A, therefore C.

The enthalpy of formation of $C_8H_{18}(g)$ is −169 kJ mol⁻¹. The enthalpy of formation of $H_2O(g)$ is −242 kJ mol⁻¹ and that for $CO(g)$ is −111 kJ mol⁻¹. Therefore the value of $\Delta H^\ominus$ for the manufacture of octane (C_8H_{18}) from gaseous H_2 and CO is −1217 kJ mol⁻¹.

We see no way of choosing between A and B as the starting sentence. Our preference for C as the conclusion is not based on formal logic, but on chemical intuition. Formal logic tells us that given any two of the three statements (A, B and C) then the third one can be determined.

Chemical intuition tells us that enthalpies of formation of many molecules (even quite complex ones like C_8H_{18}) are readily available; C_8H_{18} can be obtained in pure form and therefore its enthalpy of oxidation can be measured and hence its enthalpy of formation determined. In contrast, it is difficult to imagine how C could be determined directly since the reaction between CO and H_2 is bound to produce a range of products.

It follows that C is the most likely conclusion.

3 | All these answers, except for F, are plausible and have been postulated. As with many well known and apparently simple tests, the chemistry behind this one is complex and imperfectly understood.

5 | **(a) Our choices are A, C and D.**

Note that the question says 'which of the following statements'.

A is the assumption behind the statement that 'a molecule at the surface experiences a net attraction toward the interior of the liquid'.

C is the assumption behind the statement that 'the surface tension of a liquid can be affected by dissolved substances'.

D is an underlying assumption behind the third paragraph. If the water molecules interacted more strongly with the pin, they would wet it and so it would not float. A pin, being long and thin, has a relatively large surface area to volume ratio, and this allows the forces due to surface tension to be stronger than those due to gravity.

B and E are both true. However, in order to understand the passage it is not necessary to assume the units of surface tension or the method by which it is measured.

F is, of course, true!

(b) Since a number of books have been written on the art of scientific writing, the advice we offer in this commentary can only obviously be rudimentary.

There are two key points to bear in mind when writing for a non-scientist reader. The passage must be interesting and understandable. It must be interesting because the reader may not need to know about the topic; the incentive to read on is therefore that the topic engages the reader's interest.

Think of a real example of the kind of person you are writing for. This may be a fellow student, a friend or a relative. Plan how you can capture their interest, and think carefully about the level of knowledge you can assume they have. You need to identify the key scientific ideas you are trying to get across and the specialised language you will have to avoid (or explain). Take special care over words which are in common usage, but which have a specific meaning in chemistry (for example, 'organic') and avoid the use of jargon.

Don't be surprised if you find that you need to rewrite your passage, perhaps more than once, before you are satisfied. The best test of success is to ask a non-scientist friend to read it and give an honest opinion: Is it intersting? Is it understandable?

For this example, to capture interest you might try an approach like:

Have you ever tried to float a piece of steel on water? If it is the size of a pin you will succeed — unless you put soap in the water, in which case it will always sink. How can we explain this?

The scientific idea you need to get across concerns interaction between molecules. You may feel that 'molecule' is a familiar word to most people, but the average reader may have a very vague notion of what a molecule is, even less idea that molecules 'interact' or 'bond' or 'are attracted to' other molecules, and almost no idea that water molecules interact with each other more strongly than with molecules in air. You also have the problem of exploring why adding soap to the water makes any difference.

7 (a) Our choice is A.

You are quite likely to agree with more than one of these statements (possible even all of them). Choosing the one you agree with 'most strongly' must therefore be a matter of opinion and several of the options might be defended. Nevertheless, we take the view that A is the best answer.

We regard A as an important comment on statistics, and one which is not brought out by the passage. Statistics can give information about the proportion of people who smoke who also suffer from ill health; you can interpret this as the risk run by smokers. What statistics cannot do is to tell you whether or not a particular smoker will be one of those whose health is damaged. So you cannot know whether smoking damages *your* health (even by using statistics).

We hope no-one regards B as a statement to agree with strongly. To do so would show a lack of understanding of the nature of statistics and of risk, and also of the use of illustration in an argument.

C is true. However, we think that Dawkins is using the example of smoking to illustrate that an understanding of statistics is often important in today's world. It is a very condensed argument, with a big jump to the conclusion, but we think there are more useful comments (like A) to make about the passage.

E is a worthwhile statement for those who agree with Dawkins' final sentence. Our view is that it is nevertheless not the best critical comment on the passage.

D and G are opposed sceptical views suggesting that more statistics/statisticians either would not make a difference or possibly that they would make matters worse. We can accept that it is possible to hold any of these views, and so think that they do not contribute to the debate on the passage.

F. There is much evidence that smoking damages health: whether you consider this as 'proof' depends on the standards of proof you demand, or whether you think that proof is possible elsewhere than in mathematical and logical systems.

(b) See comments on Exercise 5.

9 **(a)** **Our choice is D** – we prefer this to C for the reasons given below.

A is a good example of an imprecise conclusion which deserves the response 'it depends'. The passage points out that a given 'amount' (whether this is in mass terms or molar terms is not specified) of methane has a bigger greenhouse effect than carbon dioxide. It does not follow that more attention should be given to methane (see C and D below).

B can scarcely be a reasonable conclusion of the passage unless coal mines, natural gas pipelines and landfill sites are negligible contributors to methane production from human activity. If the estimate of the overall error is 20%, but a major contributor is reasonably accurate, then the unknown factors must be uncertain by *more than* 20%.

C is true. However, we believe that alert scientists should be aware that emissions of methane and nitrous oxide are both increasing, and that already they are believed to make a significant contribution to the greenhouse effect. In this case we regard the lack of this information as a pity (it is not in the rest of the passage either), but we prefer to support D, given that only one choice is allowed.

D is a judgement (see C above). In our view it is one worth taking, since it may be better to take unnecessary action now rather than to wait until it is clear (with hindsight) that early action would have been effective.

E may be a part of the action in D, but we do not see how it can be singled out as a specific statement to support.

(b) See comments on Exercise 5.

11 | **Our choice is C.**

Many undergraduate texts discuss HF as a strongly hydrogen-bonded system and this leads to the widespread belief that fluorine is a good hydrogen bond acceptor. By analogy with nitrogen and oxygen and an understanding of the need for a highly electronegative small acceptor atom this view is reinforced and is generally speaking true. Thus knowing that fluorine would *not* be a good acceptor is a prerequisite for appreciating the crucial statement that difluorotoluene 'is unable to form strong hydrogen bonds'. C explains this clearly and concisely.

A is not true for fluorine on an aromatic ring (see C).

B is a false argument: in the two-dimensional diagram shown it may look plausible, but in a DNA double helix rotation or flipping of the ring is not possible.

D Thymine has an aromatic sextet of six electrons, and is therefore heteroaromatic. Heteroaromatic rings are often regarded as good analogues of aromatic systems. In this case the requirement is for two molecules of similar shape and size, and this requirement is met.

13 | **(a) Our choice is B.**

The introductory paragraphs of a scientific paper often contain a summary of the existing state of knowledge in the area as an explanation and a justification for the work to be described. The authors here are outlining the problems currently encountered in the determination of ascorbic acid in beer in order to put into context a potentially better method which they describe in the paper.

A. Even if you hold this opinion yourself there is no justification for selecting this option since there is no reason to suppose that the authors wish to make this point.

C, D and E are all stated as true in the passage. Together they justify the statement in B. But no one of them could reasonably be regarded as summarising the message of the passage.

(b) See comments on Exercise 5.

15 | **Our choice is C.**

Some support for statement A, B, and D is quoted, but our view is that the author did not regard any of these as unequivocal evidence in favour of one particular pathway.

Route b is supported by 'experiments on related systems'. Route a is supported by the supposition that it 'could proceed through the "favoured" *6-endo-trig* pathway'.

The possibility that both routes occur simultaneously is suggested by 'attack at the indole 2-position is known to *compete with* attack at the 3-position.'

We see no reason to suppose that the author believes in E.

17 | Scientists are frequently expected to make decisions or to give advice in the absence of complete information. This passage provides an example. Your answers to these questions will be determined by the view you take on the nature of proof.

(a) You need to consider how you could obtain clear proof that the virus infects the rabbit and no other species. This is a demanding criterion even if you interpret 'no other species' to mean 'no other species of mammal living in Australia'. In this case, you need to consider how you would decide whether it is ever safe to use a biological method for controlling pests.

(b) You need to consider what you would regard as 'clear proof' that the species tested were not affected. You may feel you cannot make a judgement without more information about (for example) how many animals of each species were exposed to the virus, how and over what period the animals were exposed to the virus, how long the period of incubation was.

(c) You need to consider whether carrying out 'the most comprehensive study that we know of' can reasonably be regarded as obtaining 'clear proof that (the virus) infects just one species'. If you think the two statements are not consistent, you need to consider what you would regard as clear proof (see (a)).

(d) See comments on Exercise 5.

19 | According to the first model the most important factor which limits the water-solubility of non-polar molecules (like hexane) is the unfavourable entropy term caused by the ordering of the water molecules.

The second model suggests that it is the unfavourable enthalpy term which is the most important factor.

The relative contributions of enthalpy and entropy to a process can be determined from the effect of temperature on the equilibrium constant of the process.

$$\Delta G^{\ominus} = -RT \ln K$$

$$\Delta G^{\ominus} = \Delta H^{\ominus} - T\Delta S^{\ominus}$$

In this case the equilibrium constant for the process of interest is represented by the solubility of the non-polar molecules – this represents the free energy of transfer of non-polar molecules from the pure compound into water.

Qualitatively you can see that if entropy term is unfavourable ($\Delta S^{\ominus}$ is negative) then $\Delta G^{\ominus}$ will get more positive (or less negative) as temperature increases. This would mean that solubility *decreased* as temperature increases. This is what is observed for non-polar molecules.

A quantitative experiment can be carried out by determining $\Delta H^{\ominus}$ and $\Delta S^{\ominus}$ for the transfer into water of two sparingly soluble alcohols (pentanol and butanol); the difference can be attributed to a single $-CH_2-$ group. This experiment shows that the enthalpy term is close to zero, and the entropy term is unfavourable.

Thus, as you would expect, the textbook model is supported by good evidence. It is nevertheless worth noting that unsubstantiated statements are commonly made in textbooks, and it is a worthwhile challenge to discover the evidence on which they are based.

21 **(a)** If HCN is produced, clearly oxidation has been incomplete. This occurs when the temperature is not sufficiently high (or when there is insufficient oxygen).

(b) **(i)** The temperature of an open fire is not high enough to ensure complete combustion. Although most of the products of combustion go up the chimney, there is a risk that some will flow back into the room.

(ii) Incomplete combustion will still occur, but there is a reduced risk of combustion products flowing back into the room.

(iii) The high temperatures achieved in furnaces, coupled with the fitting of after-burners, ensures that combustion is complete. Although this point is not made clear in the extract, the author clearly intends that 'solid fuel boilers' should refer to industrial boilers operating at high temperatures.

(c) See comments on Exercise 5.

23 **(a)** **(i)** You can do a 'back of the envelope' calculation to give an indication of the rate at which a tree can remove formaldehyde from the atmosphere.

The maximum rate of absorption is quoted as 137×10^{-9} dm^{-2} h^{-1} ppb^{-1}; the implication (confirmed by data in the paper) is that the rate increases linearly with the atmospheric concentrations of formaldehyde.

Consider a large tree (radius 5 m, covering a surface area of 75 m^2). The leaves have two sides, and overlap of leaves means that they cover a surface greater than the area under the canopy. An estimate of 500 m^2 seems reasonable.

A large tree will remove about 70×10^{-6} g h^{-1} ppb^{-1}. Assuming ppb is a molar ratio, 1 ppb is 10^{-9} moles in 22.4 dm^3 or about 10^{-6} g in 1 m^3 of air.

A large tree could in one hour remove the amount of formaldehyde contained in 70 m^3 of air. This estimate is not affected by the concentration of formaldehyde because the absorbance rate increases linearly with the concentration.

If the tree is 5 m high it will occupy a volume of about 700 m^3.

It will take about 10 hours for the tree to clear the atmosphere around it.

This calculation is based on a maximum rate for the Lombardy poplar, which is unlikely to be sustained over a 24 hour period, since it is affected by light intensity (as implied by the specification of light intensity used for the quoted rate).

Is this a useful rate?

(ii) At a formaldehyde concentration of 2000 ppb the trees are presumably absorbing formaldehyde at about 40 times the rate which would be obtained at the highest atmospheric concentrations (of 50 ppb). Thus the 8 hour test period is equivalent to (at least) two weeks of exposure to reasonable concentrations. Furthermore, if the trees are not damaged when absorbing at this high rate, it might be argued that they would be able to cope with a proportionally longer period at a lower absorption rate.

(b) The passage contains many terms likely to be unfamiliar to chemists (e.g. transpiration rate, stomata, rapidly metabolised). Notice that the 'back of the envelope' calculations do not depend at all on understanding these terms. It is a useful lesson to realise that the essential message of a technical report can often be grasped even though many of the terms are unfamiliar.

25 | (a) The author's line is derived from the conversion factor based on the calibration samples; it therefore passes through the origin and has a slope of 1.

(b) **Our choice is C.**

Judging whether the test data confirm that the conversion factor is valid depends very much on the two points around 60 units. In our view, it would be more convincing if there were more points in the region of 30−60 units, or (if few real samples fall in this range) more high values.

We do not like B, if only because of the philosophical point that it is not possible to *prove* that the conversion factor is valid for all samples. We would be more favourably inclined to B if it started 'The data shown on the graph *are consistent with the belief that . . .*'. However this does not really deal with the problem of most of the data lying at the bottom end of the scale.

D might be true (in the sense that it may be a better representation of a 'line of best fit'), but it is not very helpful. Redrawing the line is equivalent to recalculating the conversion factor for the new NIR method. There seems no virtue in doing this for the test data alone. If there is a case for recalculating the conversion factor, it would make more sense to base it on all the available data (calibration and test).

We regard A as unnecessarily critical. It is certainly true that the scatter of the points show that at least one (and more probably both) methods lack precision. This is quite common when complex products like these are being analysed. Whether either (or both) methods is useful depends on the precision which is required to make the measurement useful.

27 | **Our choice is C.**

We reject A and B although both could result in a curve with the shape shown. However, any competent experimentalist would guard against them. Furthermore, the graph comes from a paper published in a reputable journal. It follows that it has been accepted by at least two independent referees who would have satisfied themselves that the data were not affected by instrumental inadequacy.

D is true, but this does not explain why the fluorescence is affected by pH.

C is a perfectly plausible explanation. No details are given of the nature of the interaction between zinc ions and tyrosine. But it is possible that it depends on the tyrosine being ionised (which it will not be at low pH) and it may also require that either the amino group is positively charged or that the phenolic group is unionised.

The structure of tyrosine is shown below.

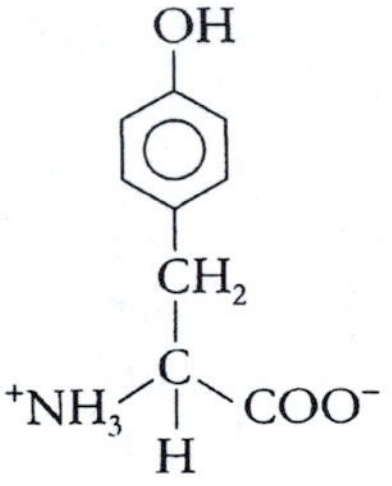

29 | **(a)** **(i)** After seven doubling times (140 mins) the mass of bacteria would increase to 128 μg.

(ii) From (i), the bacteria increase at least 10^2 times in 140 min. There are 10 periods of 140 min in 24 hours. So at this rate of doubling the bacteria will increase by 10^{20} during 24 h. Thus 1 μg will have increased to 10^8 tonnes!

(iii) Growing bacteria need warmth and food, and an environment in which their own waste products have not accumulated to an unacceptable concentration.

(b) **(i)** The population of the world is about 6×10^9.

An adult needs about 40 g protein per day, so the total daily requirement is about 240×10^9 g.

About 15% of the mass of the bacteria is protein, and so about 1.5×10^6 tonnes need to be harvested daily. The answer to (a)(ii) may convince you that this amount could easily be produced by bacteria doubling every 20 min in a shallow lake. Is such a doubling time sustainable? If it were, would there be harvesting problems?

Essex has a surface area of about $4 \times 10^9 \, m^2$. Suppose a 'shallow' lake is 0.5 m deep, the volume of the lake will be $2 \times 10^9 \, m^3$. Thus $1 \, m^3$ of lake would have to produce about 600 g bacteria per day (about $500 \, mg \, dm^{-3}$).

The bacteria need nutrients. The most significant (in terms of quantity) are a source of organic carbon (bacteria do not photosynthesise) and a suitable source of nitrogen (ammonium or nitrate ions). Bacteria are (or have been) grown as a source of animal food on methanol and ammonia (and small amounts of other salts).

In a suitably designed industrial fermenter it takes about 2 tonnes of methanol and 0.2 tonnes of ammonia to grow one tonne of bacteria.

(ii) No known animal converts food into body weight more efficiently than broiler chickens. Under intensive rearing conditions a broiler chicken consumes about 2 kg of food for every kg of weight gained. However, the calculation is not as simple as this. Only about 20% of the chicken's weight is edible meat, and over 70% of the meat is water. (Whereas the food is almost dry.) This means that the 2 kg of dry food is converted into only about 100 g of dry weight of chicken meat.

(c) See comments on Exercise 5.

31 (a) We do not know what you would regard as a 'spout'. However 10 cm³ of oil would be about 300 drops from a dropping pipette. You might regard 5 drops per second as fast enough to represent at least a steady flow.

(b) In southern UK, the total solar energy falling on the ground in a year is about $3.5 \times 10^9 \, J \, m^{-2}$ or about $0.4 \times 10^6 \, J \, m^{-2}$ in an hour. You can make any reasonable assumption about the area covered by a tree.

Photosynthesis involves using water to reduce carbohydrate according to the equation:

$$nH_2O + nCO_2 \xrightarrow{h\nu} (CH_2O)_n + nO_2$$

A maximum of 30% of visible light energy absorbed by the plant can be converted into chemical energy of carbohydrate. This is a feature of the photosynthetic pathway – it takes eight photons to reduce one molecule of CO_2. Genetic engineering cannot change this.

The product of photosynthesis is $(CH_2O)_n$ which is more highly oxidised than 'oil'. The simplest equation for the production of oil from carbohydrate (assuming a 100% efficiency of conversion, and representing 'oil' as $CH_3(CH_2)_{14}COOH$) is

$$(CH_2O)_{23} \rightarrow CH_3(CH_2)_{14}COOH + 7CO_2 + 7H_2O$$

1 g of oil releases about 50 kJ of heat when it is oxidised.

33 **(a)** If the rate of exchange is first order, then a plot of the natural logarithm of unexchanged amide hydrogens against time will produce a straight line. The data shown here do not behave like this.

(b) A simple way to test the authors' conclusion would be to assume that the two slowly exchanging groups of amide hydrogens exist and have the characteristics claimed, and then use the number and rate constants to calculate a theoretical curve. This theoretical curve could then be compared with the data for closeness of fit.

35 **(a)** This is described in the first two paragraphs of the passage. A key point is that the total triglyceride from each sample was extracted into chloroform, but only a sample of this was transferred to a vial for counting.

(b) Saponification is the process of hydrolysing triglycerides into glycerol and the salts of the fatty acids; if you were unsure about this, you could infer it from the description in the third paragraph of the passage.

(c) The sample tubes contained an aqueous solution. Water is only sparingly soluble in chloroform; the addition of methanol allows more water to form a single phase with the organic solvent, but the amount of water in the original sample means that the chloroform separates out almost quantitatively.

(d) Chloroform makes up two-thirds of the added 5 cm^3 of solvent. The washing solution is saturated with chloroform, and so the final volume of washed chloroform is 10/3 cm^3.

(e) This is a short-hand way of saying that a sample was taken from the heptane extract, transferred into a vial containing Brays scintillation fluid, and the rate of radioactive decay determined by counting the photons emitted as described in the second paragraph of the introduction to the passage.

(f) The point is not covered explicitly in this passage (or in the paper). But you can easily infer that it is determined by difference. Since the amount of radioactivity in total triglyceride and in the fatty acid part is known, then the difference between these amounts is in the glycerol.

(g) Notice that the task is concerned only with the last two paragraphs. Use your experience of following instructions in a laboratory manual to decide on a helpful format, and prepare a set of instructions you would like to have to follow.

A suitable example may look something like this:

Use a volumetric pipette to transfer 1 cm³ of the sample of chloroform extract into a counting vial and another 1 cm³ into a test tube.

Place the counting vial in an oven set at 70 °C in a fume cupboard and evaporate to dryness.* Cool. Add 10 cm³ of Brays scintillation fluid. Screw on the cap and shake gently to dissolve the triglyceride.

Add 0.5 cm³ of M KOH in ethanol to the sample in the test tube. Stopper the tube and place it in a water bath at 70 ˚C for 2 h. Remove and allow to cool. Add 0.5 cm³ of M H₂SO₄ followed by 5 cm³ heptane. Replace the stopper and shake well. Leave to extract overnight. Next day, use a volumetric pipette to transfer 3 cm³ of the heptane layer to a counting vial containing 10 cm³ of Brays scintillation fluid.

Count the samples in a scintillation counter.

* Note that this may have been a reasonable procedure in 1977 but today you would take precautions to avoid expelling chloroform into the atmosphere.

37 (a) See Exercise 35(g).

(b) The most obvious change to make would be to substitute methyl benzene (toluene) for benzene. Both have similar properties as solvents, but benzene is now regarded as too carcinogenic to be safe.

MAKING JUDGEMENTS

SOLVENTS AND SOLUBILITY

3 | This exercise invites you to consider the reasons why solubility is affected by factors such as temperature and common ions. This should encourage you to think carefully about some principles governing solubility – including the application of thermodynamics to the solution process. You should notice that some of the generalisations we tend to use (like solubility increasing with temperature) do not always hold; for example, pentanol is more soluble in cold water than in hot water (see Exercise 27 of Section 1).

You can also use this exercise to explore the precision with which you use words. For example, you might argue that you can get palmitic acid into aqueous solution by adding a base, or you might argue that the base converts the acid into the palmitate ion which is soluble, but it does not change the solubility of the acid. Consideration of the common ion effect may lead you to think about the relationship between solubility and solubility product.

4,5 | Part (a) of both these exercises draws attention to some reasonably well-defined criteria. Recrystallisation requires a solvent in which the solute is much more soluble at a high temperature than at a low temperature. It is also a requirement that any impurities should be soluble enough at low temperature for them to remain in solution; on the assumption that impurities are present in small amounts this requirement may not be particularly demanding.

The key criterion for extraction of a solute from aqueous solution is clearly that the solvent is essentially immiscible with water (see Exercise 1(c) or 5(b) for your own discussions of how 'essentially immiscible' might be interpreted). It is also essential that the solute should be measurably soluble in the solvent (but notice that it is not essential for the solute to be *more* soluble in the solvent than in water).

The real judgements required in these two exercises are raised by parts (b) or (b) and (c).

In choosing a solvent you will take into account factors such as cost, availability, safety, ease of recovery or disposal, viscosity, odour, convenience.

In deciding how many times you repeat an operation you need to decide what you are trying to achieve, and what will satisfy your objectives.

6 | See the discussion of Exercise 5(c).

REACTIONS

1 | In the context of chemistry, 'stable' is used in different ways. A key distinction is between thermodynamic and kinetic stability. But it is also crucial to consider the context. The examples given here invite you to do just this. For example, the CFCs were valued highly as refrigerants because of their stability (non-flammable and not chemically reactive), and yet when irradiated with ultraviolet light they release chlorine radicals and this has been a major contributing factor to the hole in the ozone layer. Another example, C_2H_6, can be considered from at least three different perspectives: it is stable enough to have survived underground (as natural gas) for thousands of years, however it is highly flammable (reacts readily with oxygen) and also it dissociates to methyl radicals at high temperature.

2 | ^{40}K decays at an insignificant rate on a human timescale, but not on a geological timescale; it is used to date rocks.

^{14}C is used to date artefacts on an archaeological timescale, because decay is significant over a period of 1000 years or so.

^{131}I and ^{125}I are both used in medicine but for rather different purposes partly because of their different half lives. Iodine is used to label proteins used in immunoassay (a very sensitive method for selectively detecting and measuring tiny concentrations of hormones and other substances in blood). ^{125}I is preferred to ^{131}I because it lasts longer, and therefore can be usefully stored. (It also emits lower energy γ rays and is therefore less hazardous; but this is nothing to do with the half life.)

^{131}I is usually preferred when iodine is used *in vivo* either diagnostically or therapeutically. For example, small amounts of ^{131}I may be injected in order to observe thyroid function (a properly functioning thyroid gland selectively takes up iodide). Larger amounts are used to treat hyperthyroidism, because the radiation kills some of the overactive tissue. The higher energy of I^{131} is an advantage here. The short half life means that the radioactivity does not last long enough to be a hazard to patient and families.

All these isotopes have long enough half lives for the rate of decay to be ignored over the few hours which it usually takes to determine the count rate in a batch of samples; this is not true of all radioisotopes (for example ^{18}F with a half life of 112 min).

3 | Typically, equations show what might best be described as an idealised reaction. In the examples here you should look for other possible products, for unspecified stereochemistry, for unbalanced equations (and in any such cases you should decide whether there is a reason for leaving the equation unbalanced), for examples when a symbol of state would be useful. In most of these five examples, you should have little difficulty in identifying other likely products. However, like all reactions, there will almost certainly be other products which are less easy (or impossible) to predict.

You may also wish to consider the use of $\rightleftharpoons$ and whether for any of these examples, you could justify the use of $\rightarrow$.

4 | There are many factors to consider in choosing an appropriate method. Some are general; others are specific. Will the method give you sufficient accuracy and precision for your purposes, is the method simple, quick, and safe enough for your purposes, do you have the right equipment, have you used this method before (do you feel comfortable with it)? Some specific factors brought out by these examples are whether any changes in pH or volume (or temperature) can be kept to a level which will not affect the rate, and whether a method allowing a continuous measurement *in situ* is preferred to one which involves taking samples.

5 | In most cases like this, the choice is made for convenience. Consider first how you will measure the reactant. Remember that the whole point of keeping one reactant in excess is to ensure that its concentration changes negligibly during the reaction so that any change in rate is due to the changes in concentration of the other reactant. (What does this tell you about Exercise (a)(ii)?). Your choice here is to keep OH^- in excess and to measure the change in concentration of crystal violet or to keep crystal violet in excess and measure the change in concentration of OH^-. Before deciding that OH^- is easy to measure from the change in pH, bear in mind that pH is logarithmically related to H^+. You have enough information to work out the range of concentration of crystal violet which can be conveniently measured in a spectrophotometer.

Decisions about numbers of measurements to make in Exercise (a)(iii) and about actual values of variables are almost always a compromise. The more measurements you make, the more you will improve the precision of your results, but the more costly your investigation in both time and resources. Similarly, you will want to make measurements at a wide range of values of the independent variable (in this case the concentration of the reactant which is in excess – make sure you understand why!). In practice, a number of practical constraints put limits on values which can be conveniently used; in this kind of experiment two obvious constraints are that the condition of 'reagent in excess' must not be violated, and that the reaction rate must be measurable.

6 In making a decision you need to consider the effect of your choice on the sensitivity, accuracy, and precision of your measurement. This involves thinking about what is contributing to any observed absorption.

Notice that at wavelengths below about 300 nm both the un-ionised and the ionised species absorb. It follows that in this region the observed absorption may have contributions from both species.

Notice that the two spectra cross at wavelengths just above and below 275 nm (known as the isosbestic points). At these wavelengths, any change in absorption cannot be due to a change in the state of ionisation of tyrosine. Does this mean that you can predict what will happen to the absorption if a solution of tyrosine reacts to give a product?

YIELD

1 There are two points to make about a yield of 100%. First, there is the argument that strictly speaking a 100% yield would require an infinitely negative free energy change for the reaction. For many reactions the standard free energy change is negative enough to predict a yield negligibly less than 100% starting with stoichiomeric amounts of reactants.

At a practical level, it is quite unusual to use precisely stoichiometric amounts of reactants; one will usually be added in excess in order to maximise the conversion of another into product. In such cases only the limiting reactant (at best) can be quantitatively converted into product. There are several reasons for using limiting quantities of one reactant; for example, it may be expensive or difficult to obtain, or it may be difficult to separate from product if any is left unreacted in the reaction mixture. 'Yield' is therefore often quoted with respect to one reactant.

2 Higher than expected yields should alert you to the possibility of an impure or damp product or weighing errors.

Lower than expected yields are usually caused by poor experimental technique. You should consider each step and look for possible sources of this. In the example given, low yield could result from incomplete digestion of tin, loss of iodine, timing of the reflux, loss of product through hydrolysis, incomplete washing, inefficient drying, losses during recrystallisation, and so on.

3 If the concentration of a reactant or product is changed from its equilibrium value the reaction must shift to a new equilibrium position. We can use le Chatelier's principle to predict the effect of these changes. In this example, the yield of the ester could be increased by removing the water as it was produced or by increasing the amount of acid used.

4 Whether we consider a yield to be acceptable or not depends entirely on the context. Amongst the points to consider are:

- The reasons for wanting the product. Not all products are made for future use (e.g. for sale or as the starting material for another synthesis). Sometimes the synthesis is carried out to confirm a structure or to test a theory (e.g. about reactivity or about a reaction mechanism). High yields are preferred in the former cases – especially if the reactants are expensive or difficult to remove from the reaction mixture (see Exercise 1 above), or where multi-step syntheses are involved (see Exercise 5). In the later cases, yield is usually of secondary importance to the successful synthesis of product.
- The scale of the synthesis. With large scale (industrial) production a high yield is especially important, partly because it costs a lot to waste even a tiny proportion of a very large amount of raw material and partly because of the potentially high costs of removing by-products from a reaction mixture and disposing of them. Low yields can also affect the cost of reaction vessels; more material must be processed, so the vessel must be bigger or used for longer.

- Theoretical considerations. For example, a reaction resulting in a racemic mixture will (often) result in a maximum yield of 50% of either isomer.

5–8 See commentary on Exercise 4 above. Some additional points may be helpful.

5 There are two main points to consider here. One is the overall yield; twenty successive steps each with a 95% yield will result in an overall yield of less than 40%. The other is the nature of the by-products and whether their presence can be tolerated in the final product.

6 Notice two points: both alternatives produce the same amount of *cis* isomer in the first step, but the second produces in addition an equal amount of *trans* isomer. Ideally you would start the second step with pure material.

The factors you will want to take into account are:

- The required purity of your product may have some bearing on the purity of the alcohol used in the second step.
- In the absence of any other information, it would be reasonable to assume that it would be easier to separate the *cis* isomer from by-products than from the *trans* isomer.
- For a laboratory scale preparation you may prefer to have the opportunity to try *both* methods of making the chloride, or you may prefer to concentrate on one process.
- For a large-scale production: it is much more important to minimise the waste (see commentary on Exercise 4 above); some separation procedures (like fractional distillation) are not difficult, but others (like chromatography) are very problematical; it may even be worth spending resources to develop a method for separating the *cis* and *trans* isomers if one has not been published.

7 The mechanism of the reaction is not affected by the yield, and so in this case the yield does not matter at all – provided that there is enough pure product for the distribution of label in it to be determined. Indeed, in other circumstances, you could imagine wanting to determine the mechanism for the formation of a compound formed in low yield as a by-product of another reaction; this might be the first step towards working out conditions likely to improve the yield.

In this example, you could actually analyse either II or III. It seems likely that III is formed from II; thus 76% of I is converted into II, although two-thirds of it is then modified further. However, even if the nitrogen in the five-membered ring is derivatised first, it should not affect the cyclisation reaction being studied.

8 | See Exercise 5 above. Each step in the synthesis would usually be based on known reactions. For each of these you would need to be convinced that the reaction had proceeded according to plan and given the expected product. What minimum yield would you accept for each step? What average yield for each step would lead to an overall yield of 1%?

When proving a structure, it is more important to be confident that each step leads to the expected product than to obtain a high yield.

ACCURACY AND PRECISION

The difference in meaning of accuracy and precision is explained in the introduction to this section. Precision without accuracy is all too easy to achieve: for example, you may make many repeat readings of the pH of a solution and find that each reading is virtually the same (high precision), but if the pH meter is wrongly calibrated it will be highly inaccurate. Accuracy without precision is probably less common, but you can imagine a method subject to high experimental error which will give an accurate mean value providing that sufficient readings are taken. If you make a single measurement using a method of low precision, can you get an accurate result?

1 | You need to consider here which of the suggested actions can give information about systematic error. See introduction above.

2 | You need to think rather carefully about the relationship between accuracy and precision in this question; you also have to apply your experience with laboratory work (or use common sense).

Would you expect a top pan balance to be accurately calibrated? If the balance is placed on an ordinary laboratory bench, do you expect a reading to be reproducible? Does the accuracy of the weighing determine

the accuracy with which you know the concentration of the resulting solution?

3 A balance will nearly always give greater accuracy and precision than can be obtained by measuring a volume. The problem with diethyl ether is its volatility. The confidence you have in the value you obtain will reflect your confidence that you have overcome this problem.

4 (a) You need to think about the purposes for which you use these pieces of equipment. Measuring cylinders and graduated pipettes are not expected to be as accurate as volumetric glassware. When measuring volumes or making up solutions the greatest accuracy you can achieve is determined by the smallest volume you can reasonably measure. In routine analysis this is roughly half a drop (about 0.02 cm^3). Is it worth calibrating equipment any more closely than that to the correct value?

(b) What is the relationship between the precision of the calibration method and the closeness of the calibration value to the correct value? If anyone needs to know about the precision of the method, is it the user, or the calibrator, or both?

(c) You could weigh the amount of liquid contained in each. How would you do this? What liquid would you use? Would you use the same procedure to weigh the amount of liquid in a volumetric flask and a volumetric pipette?

5 You need to consider the expected variation from each of the methods. Since you are given no information about the precision of either, you are entitled to ask whether 99.5% can be distinguished from 99.8% or whether the two results are not significantly different. You might also ask whether the values are sufficiently different from 100% to justify concluding that the amine is impure.

If you conclude that they are (or may be) different, you need to consider what is being measured in each case and what impurities might give rise to an error. For example, an amine contaminated with another amine would give a different titre from an amine equally contaminated, but with a neutral (or acidic) substance. With chromatographic procedures, some impurities may not show up at all.

6 | The measurement of the concentration of alcohol in blood is subject to experimental error. Therefore, there is a possibility that a concentration of less than 80 mg per 100 cm^3 will result in a measured value of more than 80 (by the same reasoning, a real concentration of more than 80 may give a reading of less than 80). You need to consider how far above 80 the mean measured value has to be before you can be confident that it does not arise from a legal concentration as a result of experimental error. The standard deviation for the determination is about 2 mg per 100 cm^3, and so the standard error of a mean of four values is about 1 mg per 100 cm^3.

7 | A signal-to-noise level in excess of 2 used to be widely accepted as significant, but now 3 is commonly demanded in practice. A minimum of 10 is usual for quantification; any measurement at this level can be no better than ±10%! What suffices scientifically really depends on circumstances, the nature of the noise and drift, the acceptable certainty and (increasingly) on the pre-processing which occurs in modern instrumentation. There are many ways of measuring noise: peak-to-peak over a given period, maximum excursion from baseline, a root mean square measure, or averaging of noise over several runs. Estimation of signal and estimation of noise both require judgement.

8 | There are at least two issues to consider here. First, on what basis can the analyst claim to know that the probability of obtaining a false positive is 1 in 5000? You might suggest that it is based on testing a very large number of samples which are known to be free from oestrogen and finding that 1 in 5000 gave a positive result; it would be hard to justify this value if less than 100,000 samples had been measured, and this seems unlikely. Whatever method has been used, the value will be an estimate, and the precision of the estimate is likely to be unknown.

The second is whether someone should be convicted if there is a 1 in 5000 chance of a mistake; at what level of doubt do you give the benefit of the doubt? There is a roughly 1 in 5000 chance of tossing 12 successive heads (or tails) with an unbiased coin. If you tossed 12 successive heads would you conclude that you had a weighted coin?

Most people would feel that the more serious the charge the lower should be the probability of obtaining a wrong analytical result by chance.

Against this, one should remember that a conviction rarely depends on a single analytical result, but on several pieces of evidence.

EQUILIBRIA

1 You need to consider each part of this question from more than one viewpoint. Is there a difference between pure water, water as a solvent, and water as a solute? Remember that on thermodynamic grounds you would predict that a dissociating system will become more dissociated as it is diluted; is this a fair analogy for diluting water with propanone? Acids do not have to be dissolved in water. An equilibrium constant is determined in relation to standard conditions; the conditions in the atmosphere or inside living cells differ markedly from standard conditions. Whether or not you regard a reaction as going to completion depends on the context. For example, the concentration of H^+ resulting from the dissociation of water is generally regarded as highly significant; whereas the amount of undissociated HCl in a dilute aqueous solution can usually be ignored. Some relevant points are also made in the commentaries on Section 4.5.

2,3 You need to consider what you mean by 'complete', whether there is a difference between a strong acid and a strongly acidic solution, and whether pH reflects concentration or activity.

4 See commentaries on Exercises 2 and 3 above, but also decide on the pH range over which you can ignore the proportion of HA^+ and of A^-, and consider the pH range over which any substance can act as an effective pH buffer (think of this range relative to the pKa of that substance).

5 In 1 mol dm^{-3} H_2SO_4 you might expect organic acids to be fully protonated, in which case there could be no sodium salt. This cannot be true if the sodium salt crystallises in preference to the acid; what sort of pK does this imply?

6 You can use the solubility products to calculate the concentration of both Ba^{2+} and Sr^{2+} which will precipitate at any concentration of chromate. Since both metal ions are in the same solution it is easy to show that in theory they will co-precipitate when the metal ions are present at a concentration ratio equal to the ratio of the solubility products. The question is whether the system will behave according to pure theory, or whether Sr^{2+} may be carried down with some barium chromate.

To use this procedure to purify Ba^{2+} or Sr^{2+} you have to decide how much impurity you will tolerate.

7 You can answer this question quantitatively by using the equilibrium constants. But you can avoid a tedious calculation by the intelligent application of le Chatelier's principle. If you compress a sample of NO_2 to one-tenth of its volume, why would you not expect the pressure to go up ten times?

8 As with Exercise 7 you can answer this question by calculation, or you can do it much more easily by applying the principle that when you dilute a solution containing associating species there is a decrease in the equilibrium ratio of complex to dissociated components.

9 The key point here is to recognise that you do not need to obtain precise numerical answers, but you can make simplifying assumptions which make the arithmetic easy whilst still giving useful information.

Remember that the equilibrium expression gives:

$$\frac{[complex]}{[EDTA][Ni^{2+}]} = 4 \times 10^{18}$$

Adding a stoichiometric amount of EDTA should complex essentially all the Ni^{2+} and leave equal (but tiny) concentrations of free Ni^{2+} and EDTA; the value can be estimated from the stability constant given. A colour change will be seen at this point if the stability constant of the murexide–Ni^{2+} complex is low enough for a significant proportion of murexide to be uncomplexed at this concentration of free Ni^{2+}. A colour change will occur too soon if the stability constant for the murexide complex is too low.

To tackle Exercise (b) you have to decide what concentration of free Ni^{2+} you would describe as 'effectively zero'; would 10^{-24} mol dm^{-3} be appropriate? This is less than one ion per litre; you may think it unnecessarily low. You can put your chosen value, together with the total concentration of Ni^{2+} (which is completely complexed) into the equilibrium equation and calculate the concentration of free EDTA. Could you make a solution of this concentration?

This is the only chapter for which we can provide unequivocal solutions to the exercises.

Where no reference to a paper is given, the answers to the questions can be found in the starting reference.

References following a particular answer provide a source for that answer and for all subsequent answers until a new reference is provided.

1 (a) Self assembled monolayers are monolayers of organic molecules which form spontaneously on a solid substrate by adsorption from solution.

(b) Images were collected using LFM, lateral force microscopy, and showed a close-packed structure with homogeneity extending to at least a few micrometres. (*Langmuir*, 1994, **10**, 2241)

(c) Formation proceeds by nucleation growth and coalescence of sub-monolayer islands.

(d) The monolayer island growth is proportional to:

concentration $\times$ time$^{1/2}$

3 (a) Supercritical fluid – a substance above its critical point, which is the highest temperature and pressure at which its vapour/liquid equilibrium can exist.

(b) RESS – rapid expansion of supercritical fluid solutions.

(c) Product morphologies include microfibres, microspheres, hollow microfibres, microencapsulations, etc.

(d) Gas anti-solvent relies on the high solubility of supercritical fluids in liquids at low pressure, causing the solvent strength of the liquid to decrease and selectively precipitate solutes.

(e) Precipitation with compressed anti-solvent involves spraying a liquid solution containing the desired compound into the supercritical fluid. Droplets dissolve in the supercritical fluid inducing precipitation.

(f) Unique and potentially advantageous characteristics of supercritical fluid solvents are their density, polarity, viscosity, diffusivity and overall solvent strength which can be dramatically varied by relatively small changes in the pressure and/or temperature.

(g) Critical temp = 31 °C
Critical pressure = 72.9 atm

(h) Extractions from, for example, bark, hops, spices, oil seeds, fish oils, roots, beans, fruits.

Solvent for cleaning, drying.

Synthesis of formic acid, catalytic hydrogenation.

5 **(a)** Ordinary distilled water was redistilled from alkaline permanganate and was then passed through columns of Amberlite IR-120 and Dowex 50 ion exchange resins in H^+ form. Afterwards it was boiled with peroxodisulfate (1 g dm^{-3}) for ca. 30 minutes, neutralised with purified sodium hydroxide solution and distilled from a pyrex apparatus. (*J. Chem. Soc., Dalton. Trans.*, 1982, 237)

(b) pH 11–12.

(c) Various glasses, pyrex, varnished surfaces, metal containers, etc. (*J. Phys. Chem.*, 1943, **47**, 260).

(d) Addition of a chelating agent will remove metal ions. Compounds used H4edta, H5dtps, H4dodta, H3nta, H4cdta.

7 **(a)** Add the catalyst to hot sulfuric acid and stir for 15 min. Titrate with 0.03 mol dm^{-3} ferrous ammonium sulfate. (*J. Catal.*, 1982, **76**, 17)

(b) The diagram is given in *J. Chem. Soc., Faraday Trans.*, 1993, **89**, 843.

(c) ca. 380 °C (*J. Catal.*, 1975, **37**, 424)

9 **(a)**

(b) 10% (*J. Am. Chem. Soc.*, 1971, **93**, 1259)

(c) 66% (*J. Am. Chem. Soc.*, 1993, **115**, 7246)

11 **(a)** Physically separated reactants.

(b) Products would be impossible to separate.

(c) Metathesis.

13 **(a)** Dissolve 0.01 mol of acid, 0.011 mol of dicyclohexylcarbodiimide, 0.011 mol of the alcohol and 0.001 mol of 4-pyrrolidinopyridine in ether or dichloromethane (25–30 cm³) and allow to stand at room temperature until esterification is complete. Filter; wash with water,

5% acetic acid, and again with water and dry over magnesium sulfate. Evaporate to dryness to give the ester.

(b) Neutral conditions.

(c) NMR.

15 **(a)** $SbF_5 > AsF_5 > NbF_5 > PF_5$
Ce^{3+} and Ce^{4+} (*J. Inorg. Nucl. Chem.*, 1957, **5**, 57)

(b) A mixture of *m*- and *p*-xylene was treated with HF plus the metal salt. In this medium a fluoride may act as a Lewis acid and coordinate with a fluoride ion. The proton released then adds preferentially to *m*-xylene.
    The acidity of the fluorides was attributed to the high positive valence of the metal and its ability to form covalent bonds with fluorine. $TaF_5 > NbF_5 > PF_5$ (*J. Am. Chem. Soc.*, 1956, **78**, 3009)

(c) $PF_5 < NbF_5 < TaF_5 < AsF_5 < SbF_5$

(d) $SbF_5 + 2HF \rightleftharpoons H_2F^+ + SbF_6$
$2AsF_5 + 2HF \rightleftharpoons H_2F^+ + As_2F_{11}^-$

(e) $2SbF_6^- + H_2F^+ \rightleftharpoons Sb_2F_{11}^- + 2HF$
SbF_4SO_3F in HF (*J. Chem. Soc.*, 1966, **118**, 1170)

17 **(a)** The Huang–Minlon reaction is a modified Wolff Kishner reaction for the reduction of keto acids and ketones.

(b) Your method should be based upon the following:
Reflux the compound in ethylene glycol with 85% hydrazine hydrate and 3 equivalents of NaOH or KOH for 1 hour. Distil off water and hydrazine hydrate to raise the temperature to 180 °C and reflux for a further 3 hours. Isolate compound on cooling. (*J. Am. Chem. Soc.*, 1946, **68**, 2487)

19 Potentiometric titration with $KMnO_4$ using Pt/Ag/AgCl electrode to determine amount of V^{4+} then volumetric titration with Fe^{2+} to determine V^{5+} using sodium diphenylamine sulfonate indicator. (*J. Chem. Soc. Faraday Trans.*, 1994, **90**, 2981)

21 **(a)** Between 0.005 and 0.01 M solution of the steroid in acetic acid is shaken with zinc dust. Shake for 45 minutes. Filter, wash with acetic acid. Add ice and neutralise. Extract with ether. (*J. Chem. Soc.*, 1959, 2502)

23 (a) Chemical vapour deposition, 1.0 eV. (*J. Phys. Chem.*, 1992, **96**, 2691)

(b) 1.7 eV. (*J. Phys. Chem.*, 1992, **96**, 2691)

(c) Laser pyrolysis. (*J. Am. Chem. Soc.*, 1984, **106**, 390)

25 (a) Kroto *et al.*, *Nature*, 1985, **318**, 162.
The method used was laser vaporisation of graphite.

(b) Previous studies gave conflicting results; hence this study.

(c) Maity *et al.* used pulse radiolysis and γ radiolysis.
Dimitrijev and Kamat used pulse radiolysis and laser flash photolysis.

(d) Maity *et al.* 50×10^{-6} mol dm^{-3} 32 J kg^{-1}
Dimitrijev 3.25×10^{-6} mol dm^{-3} 4.20 J kg^{-1}
(note that 1 Gy = 1 J kg^{-1})

(e) Both are fairly stable towards radiolysis. Both sets of authors postulate a triplet as the excited state.

(f) C_{70} is more stable than C_{60} but both are more stable in benzene than in cyclohexane.

27 Dissolve $PdCl_2$ in minimum 2,4,6-$Me_3C_6H_2NC$ with heating. Filter then cool to give precipitate. Dry over calcium chloride and collect crystals. (*Inorg. Chem.*, 1977, **16**, 778, then to *J. Am. Chem. Soc.*, 1938, **60**, 882)

INDEX

This index is in three parts, reflecting different ways in which the book might be used. Each entry identifies relevant exercises by page number and exercise number.

- The **keyword** index presents conventional entries which identify exercises dealing with a particular topic. Many exercises are designed to illustrate principles, and use specific examples to do so. We have preferred to index keywords associated with the principles, and have often omitted references to the examples.
- The **sub-discipline** index groups the exercises according to their traditional sub-discipline of chemistry. Exercises have been classed as 'general' either because they do not fit comfortably within a sub-discipline or because the underlying chemistry is so basic that it is properly referred to as 'general chemistry'. Many exercises have been classed under more than one sub-discipline; we accept that not everyone will agree with our classification.
- The **methodology** index is most in keeping with the concepts underlying this book. We want chemistry students to learn to think – and thinking is not a one-dimensional process. The entries in the methodology index identify different aspects of the thinking process, and classify relevant exercises.

KEYWORD

SUB-DISCIPLINE

METHODOLOGY